SpringerBriefs in Space Development

Series Editor: Joseph N. Pelton

For further volumes:
http://www.springer.com/series/10058

This Springer book is published in collaboration with the International Space University. At its central campus in Strasbourg, France, and at various locations around the world, the ISU provides graduate-level training to the future leaders of the global space community. The university offers a two-month Space Studies Program, a five-week Southern Hemisphere Program, a one-year Executive MBA and a one-year Masters program related to space science, space engineering, systems engineering, space policy and law, business and management, and space and society.

These programs give international graduate students and young space professionals the opportunity to learn while solving complex problems in an intercultural environment. Since its founding in 1987, the International Space University has graduated more than 3,000 students from 100 countries, creating an international network of professionals and leaders. ISU faculty and lecturers from around the world have published hundreds of books and articles on space exploration, applications, science and development.

Joseph N. Pelton

Satellite Communications

Joseph N. Pelton
Space & Adv. Comm. Research Inst.
George Washington University
Arlington, VA 22207, USA
joepelton@verizon.net

ISSN 2191-8171   e-ISSN 2191-818X
ISBN 978-1-4614-1993-8   e-ISBN 978-1-4614-1994-5
DOI 10.1007/978-1-4614-1994-5
Springer New York Dordrecht Heidelberg London

Library of Congress Control Number: 2011941442

© Joseph N. Pelton 2012
All rights reserved. This work may not be translated or copied in whole or in part without the written permission of the publisher (Springer Science+Business Media, LLC, 233 Spring Street, New York, NY 10013, USA), except for brief excerpts in connection with reviews or scholarly analysis. Use in connection with any form of information storage and retrieval, electronic adaptation, computer software, or by similar or dissimilar methodology now known or hereafter developed is forbidden.
The use in this publication of trade names, trademarks, service marks, and similar terms, even if they are not identified as such, is not to be taken as an expression of opinion as to whether or not they are subject to proprietary rights.

Printed on acid-free paper

Springer is part of Springer Science+Business Media (www.springer.com)

Contents

Introduction to Satellite Communications

1

Why Write a "Short Book" on Satellite Communications?

First of all, satellite communications are vital to our planet's operation. Did you ever wonder how television programs get to your television set? Or perhaps considered where signals go when you stick a credit card into a gas pump when you buy some gasoline or petrol? Chances are when you see a news item from around the world – even if you watch cable television – that the signal went at least part of the way via satellite. Over 12,000 satellite television channels now routinely fly some 50,000 miles through outer space before they reach a television set or wireless telephone or cable head end. Satellites have changed how interconnected the world is today. Satellites have enabled the Internet to reach countries around the world. Satellites allow better education, better health care, more news, more and better airline connections and, in short, a more vital and democratic world. If you have ever wondered how satellites work, what orbits they fly in, or what services they provide to a contemporary world then this book is for you.

This book provides a quick overview of the essence of satellite communications. It provides all of the key information and vital trend lines in a nutshell. There are many books about the field of satellite communications. Some are written for engineers. Others specialize with regard to various satellite communications services such as satellite broadcasting, mobile satellite communications, or fixed satellite services. Some are written from a legal, regulatory, business or financial service point of view. These are useful books and references. Some of the best and most current books you will find referenced here. The point of this book, though, is to let you quickly peer into the world of satellite communications and within a few pages see just how important this technology has become and how pervasive it is around the world.

The key point of this book is that it is brief and what you might call "meaty." This book is written for the busy person who needs to know the essence of the field of satellite communications. It provides an overview of the technology – not for the engineer, but for those in other technical or scientific fields or students. We cover the useful

J.N. Pelton, *Satellite Communications*, SpringerBriefs in Space Development,
DOI 10.1007/978-1-4614-1994-5_1, © Joseph N. Pelton 2012

orbits, the services, the business and financial aspects as well as legal, regulatory and trade issues. Most books are highly specialized. They drill down in a very, very focused way – like an oil rig seeking a target far below the surface. These books can be quite deep. Unless you are interested in a particular narrow range of minutia such a book can end up being not only incredibly dull but worthless to the non specialist.

The scope of this book is broad. Its mission statement is to give you the whole enchilada. It is written in clear expository style for someone who needs all the key information about every aspect of satellite communications in a single place for quick reference. In a three-hour read you can acquire a basic knowledge of the entire field of satellite communications.

Here you can not only learn how the industry works but also where it is headed. This book provides key information on the many aspects of the satellite world. If you are interested in pursuing a career in satellites or already involved in some way in the satellite industry, then this book is perfect for you.

Satellite Communications is structured so that one can find needed information quickly and easily. Chapter 2 jumps right into satellite orbits, communications satellite services and the various types of systems and operators. This is followed by an overview of satellite communications technology and operation. It covers the key steps from the satellite's design through engineering, manufacture, testing, deployment, operation, sparing, and end of life disposal. Chapter 4 covers Earth stations, ground antennas and satellite user devices, and Chapter 5 addresses launch arrangements and risk mitigation. These first five chapters focus on today's technology and trends.

The next few chapters cover essentially everything else. Thus there are chapters on business, marketing and financial issues as well as a chapter the legal, regulatory and trade environment. These chapters explain the real world of the satellite communications industry as it exists today. Here you can also examine future satellite market trends. One of the fascinating aspects is to learn how, on one hand, satellites compete with fiber optics and terrestrial communications systems, and yet there are also significant cooperative roles as well. Terrestrial communications systems now play the predominant role in global communications, but satellites still have clearly defined supplementary market roles. The final chapter serves as an executive summary that also provides insight into the prospects for future satellite-based services. You can read this book in any order that you wish, although it is short enough that a complete read through from start to finish is recommended.

What Is the Best Way to Grasp the Field of Satellite Communications?

The field of satellite communications is not the easiest one to quickly grasp for several reasons. First, there are no two ways about it – the field of satellite communications is fairly technical. This does not mean you have to earn an engineering degree to understand the field. The most basic technical dimensions and prime concepts on which the satellites and the receiving and transmitting antennas operate are actually pretty straightforward. Accordingly this "Brief" contains no formulas or detailed engineering text, with the exception of a few endnotes for those who might wish to have the more complete technical lowdown.

Secondly, satellite communications is an interdisciplinary field and so this book covers a lot of territory. It addresses satellite design and engineering, rocket launchers, a variety of communications services, plus the business, finance, legal and regulatory matters.

Thirdly, there are many players in the field that will be discussed as well. As the satellite industry has evolved over time there has been a steady expansion of various types of satellite services. There has also evolved a range of governmental and international regulations that seek to control undesirable practices and create a framework for additional allocations of radio frequencies on which the satellite industry strongly depends. This has seen a process whereby satellite communications have always tended to move upward to higher and higher frequencies with smaller and smaller wavelengths. We have also seen the continual evolution of a globally competitive market structure that extends from local to regional to worldwide satellite operators. Along these commercial market trends we have seen the parallel development of more and more sophisticated military and defense-related satellite networks.

There are literally thousands of companies, governmental agencies, and international entities that are involved in the field. *Satellite Communications* attempts to hit the highlights and show how the overall industry interconnects. We will not get bogged down in detail but will show how the satellite industry works from a higher level – the 40,000 foot elevation overview.

One of the important stories, however, is that of digital services over satellites and especially what has been done to make Internet-related services via satellite more efficient. This is particularly important because the Internet was designed for terrestrial networks and not satellites. A great deal of effort has been undertaken over the past two decades to improve how Internet Protocol (IP) services operate using satellites. Another story that can be summarized is that of the pros and cons of on-board processing in satellite networks. The technical details are considered vital to the engineers who design satellite systems, but again we can address these issues at a high level. Thus it is possible to skip over such technical complexities as digital encoding and processing and instead provide the big picture within one condensed and we hope readable book. This is what this "quick read" book is all about.

Orbits, Services and Systems

2

Satellite Orbits

In the 17th century Galileo Galilei (1564-1642) first observed the moons of Jupiter through his handcrafted telescope. These observations ultimately led to his realization that not only did planets encircle the Sun, but that moons could and indeed did encircle planets. To the moon of Jupiter he applied the Latin word *satelles*, which means "servant." He concluded that somehow a planet 'commanded' its moons to remain in their constant orbits just as a master or mistress commanded the actions of a servant. Today this remains an apt term in that artificial satellites do the bidding of the scientists and engineers that design satellites for telecommunications, broadcasting, remote sensing, space navigation, meteorology, geodetics or scientific exploration.[1]

A century later Sir Isaac Newton not only discovered gravity but understood how gravity commanded the orbits of satellites. In his book *Philosophiae Natralis Principia Mathematica*, and known as simply the *Principia*, Newton explains the workings of gravity in terms of the "falling" of objects but also applied it to the planetary motions described by Johannes Kepler some years before. Newton's famous book even included an iconic illustration that showed how a cannon that shot high enough and fast enough could hurl an object into Earth orbit. Thus we have known for three centuries how artificial satellites could be launched into orbit to carry out various tasks.[2]

Ironically, it was the world of early science fiction that gave us our first vision of how various types of application satellites might be launched into Earth orbit and then carry out various missions. Although there are many examples from

---

[1] Joseph N. Pelton and Scott Madry, "Satellites in the Service of Humanity," Chapter 6, in Joseph N. Pelton and Angela Bukley, editors, *The Farthest Shore: A 21st Century Guide to Space* (2009) Apogee Books, Burlington, Ontario, Canada. p. 181–82.

[2] *Ibid*, Angie Bukley et al, "Space Missions and Programs: Why and How We Go to Space," Chapter 8, p. 259–60

J.N. Pelton, *Satellite Communications*, SpringerBriefs in Space Development, DOI 10.1007/978-1-4614-1994-5_2, © Joseph N. Pelton 2012

H. G. Wells, Jules Verne and others, one of the most striking first such images was provide by Edward Everett Hale in his book *The Brick Moon* (1869). Hale foresaw an artificial moon being launched into polar orbit to carry out weather and Earth observation and to act as a communications device. This turned out to be an amazing precursor prediction of what would happen more than a century later. Of course the artificial satellites of today are not built of bricks. Neither do polar orbiting satellites have a crew of 17 men and two women on board nor do they use Morse code to transmit messages. Still Hale's book was a remarkably innovative concept that helped to lead the way forward.[3]

The key to all communications satellites being able to operate successfully is to place them in an appropriate orbit targeted to providing one or more useful services. Today there are three main groups of orbits used by telecommunications. These are the geosynchronous Earth orbit, or GEO (also known as the Clarke orbit in honor of Arthur C. Clarke, who first wrote about using this unique orbit for satellite communications in 1945); medium Earth orbit, or MEO; and low Earth orbit, or LEO. Although there are other orbits that can be used for satellite applications that will be described briefly, most communications satellites are in GEO. There are, however, also a number of LEO and MEO constellations that are used for communications as well as other practical purposes such as space navigation, remote sensing, and meteorological observation.

When satellites are launched from Earth they tend to go initially into an elliptical orbit. These are just like the elliptical orbits the planets follow around the Sun when one charts the geography of our Solar System. The Sun, as explained by Kepler, is not found at the center of a series of concentric orbital circles. Rather the Sun can be found in one of the two foci found in every ellipse because a circular orbit is a very unique and special case of an ellipse when the two foci exactly overlap. A satellite launch thus typically sends the spacecraft into such an orbit with an apogee (or high point) and a perigee (or low point). This is shown in Fig. 2.1 below.

The initial launch will send the spacecraft into an elliptical orbit. Some of these orbits can be suitable for LEO or MEO constellations. If the launch, however, is for a satellite intended for GEO then the initial orbit will be a very highly elliptical orbit, called a transfer orbit, with an appropriate very high apogee (35,780 km) so that the spacecraft can later be pushed into a circular GEO. In the early days what was called an apogee kick motor would be fired at apogee to circularize the orbit. Today, a last stage rocket firing or on-board thrusters will divert the satellite to a circular GEO.

LEO Constellations

LEO altitudes are typically in the range of about 500 to 1,200 km. These orbital altitudes are essentially below the lowest of the Van Allen high radiation belts and

---

[3]Ibid, Peter Diamandis, Robert Richards and Joseph Pelton, "The Future of Space," Chapter 4, p. 118.

Fig. 2.1 Natural elliptical
orbital characteristics with a
typical satellite launch

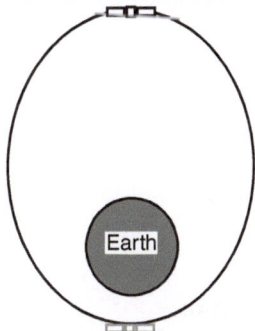

The apogee is the highest point in the orbit where the
satellite is moving at the slowest velocity.

High point - apogee: satellite is going very slow

Earth

Low point - perigee: satellite is going very fast

The Perigee represents the lowest point of the orbit
where the satellite has maximum velocity.

require a constellation of satellites of between 40 and 80 spacecraft in order to provide global coverage. One can also design an LEO system to provide coverage for only a portion of the globe such as between 65 degrees latitude North and 65 degrees latitude South. Such a constellation can "see" 99% of the world's population. (Penguins do not represent a useful client base.) If one looks at Fig. 2.3, which shows an MEO constellation, it should be clear that quite a few more satellites located at much lower altitudes would be needed to achieve complete coverage of Earth. If this is not clear shine a flashlight on a globe and see how the illumination area grows or shrinks as you move the light away from the globe or bring it back much closer.

In essence, the lower the low Earth orbit, the more satellites are needed to complete the constellations' total coverage. If you are in a balloon, for instance, the higher you are the farther you can see in all directions.

The positive tradeoff is that the flux density (or irradiated power) is higher within the much smaller viewing or "catchment area" on Earth's surface. Obviously if the satellite transmission path is much shorter the so-called path loss, or spreading out of the transmitted power, is also much less. Clearly one needs a larger "swarm" (or more precisely a constellation) of satellites to truly cover Earth's surface completely, since each LEO satellite can only illuminate a much smaller area. Enthusiasm for low Earth orbit satellite constellations for mobile communications and data relay services diminished after the bankruptcies of the Iridium, Globalstar and ICO global mobile satellite systems in the 1990s. This was further compounded by the later financial difficulties with the Orbcomm data relay system. After financial restructure,

Fig. 2.2 One of the
Orbcomm satellites launched
into a low Earth orbit
constellation. (Graphic
courtesy of Orbcomm.)

Fig. 2.3 The NAVSTAR
GPS satellite constellation for
space navigation

however, all of these systems except ICO are now operating successfully and are
technically and operationally viable. (See Fig. 2.2.)

MEO Constellations

In the case of MEO satellites the constellation of from 8 to about 24 satellites is
configured in orbits that are typically 10,000 to 20,000 km above Earth's surface.
Because the satellites are higher, fewer satellites are need to cover Earth, but the
path loss due to the spreading of the antenna beams means the flux density of the
beams is less when they reach the ground. The number of satellites in the constellation depends on not only the altitude but also the particular mission the satellites are
designed to perform. Space navigation satellites such as the NAVSTAR Global
Positioning Systems (GPS) for instance requires that a user accesses four or more
satellites to get an accurate fix on location. Thus this constellation, although in a
relatively high orbit, still has some 24 to 27 operational satellites in order that multiple satellites can be seen at the same time. (See Fig. 2.3.)

GEO Communications Satellites

In the case of GEO, the satellite is first launched into a highly elliptical (cigar-shaped) transfer orbit where the perigee (low point) is only a few hundred km in altitude, but the apogee (high point) is nearly 36,000 km. The satellite remains in this transfer orbit until an appropriate apogee is close to the desired longitudinal location with respect to Earth's equator. At this stage either an "apogee kick motor"(in the early days) or the last stage engine of the launch system or apogee motor pushes the satellite from transfer orbit into a new perfectly shaped circular orbit that allows the satellite to move around the planet exactly once every 23 hours and 56 minutes. In this very special orbit the satellite appears to remain exactly stationary with regard to Earth below. (Note: What seems to be the "missing 4 minutes in a day" actually is not missing at all. Earth travels around the Sun every 365.25 days, and this means it travels 4 minutes worth of its annual revolution around the Sun every day).

Again there are tradeoffs to consider. Only three GEO satellites are needed to cover the planet except for the most extreme polar cap regions. The great altitude means that the flux density of the beams is much less than for LEO or MEO systems because the "path loss" (i.e., the loss of signal strength equivalent to the spreading of the beam from the satellite's antenna in its journey back to Earth) is much greater for the GEO than for the lower orbits. In the case of the GEO satellite, the spacecraft is almost one-tenth of the way to the Moon. Even with a very highly focused beam, the signal spreads out greatly by the time it reaches Earth. This spreading out of the beam over its transmission distance is called "path loss."

The enormously important advantage of the GEO or Clarke orbit satellite is that ground Earth stations, very small aperture terminals (VSATs) and various forms of antennas, from larger Tracking, Telemetry and Command (TT&C) units to small micro-terminals, do not have to "track" or continuously move to keep connected. Once one orients a ground station, particularly a small aperture VSAT or a direct broadcast receiving station, it stays continuously connected because the satellite appears to remain constantly fixed overhead.

There are two terms often used with regard to a GEO satellite – one is "geostationary" and the other is "geosynchronous." A geostationary satellite would be one that remains constantly at the same longitude all the time and also remains exactly in the equatorial plane at 0 degrees latitude. Such a perfectly geostationary orbit is, in fact, almost impossible to achieve in the real world because of Earth's orbital mechanics. This is because Earth is first of all not a perfect and homogenously composed sphere of constant density. An even more important factor is that the gravitational pulls of the Moon and Sun are constantly changing forces, tugging at GEO satellites pulling them either east or west in longitude or even more strongly dragging them north or south to higher "inclination." Inclination means the degree of elevation above or below the equatorial plane.

The north and south gravitational pulls of the Sun and Moon are the strongest and indeed are ten times greater than the east and west longitudinal forces. The satellite thus has to be managed through "station-keeping" to keep it inside of a box

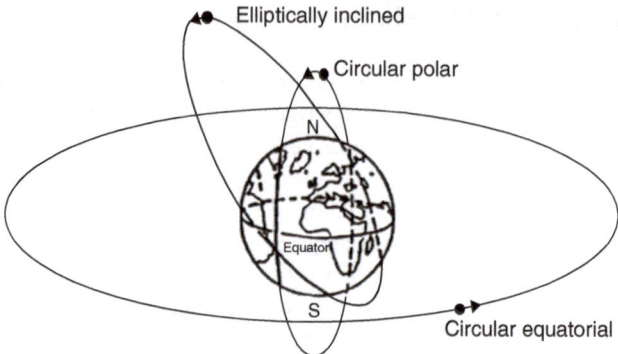

Fig. 2.4 Comparing satellite orbits (not to scale). (Graphic courtesy of author.)

by the firing of rocket thrusters to keep it "geosynchronous" or at the right speed to complete the one complete revolution per sidereal or "solar" day. Most geosynchronous satellites have some inclination build up and go up and down a small bit from the equator each day and thus are not truly geostationary, especially in terms of north and south latitude excursions. Unless you are an astrodynamics engineer you can forget about the distinction.

What's important to visualize is the extent of the major geographic separation in altitudes that the various orbits represent. A GEO satellite is about 3 times further out from Earth than a MEO. The GEO (which is the outer orbit shown below) is, in fact, about one-tenth of the way to the Moon. A GEO can be up to 40 times further away than a LEO satellite. Figure 2.4 below is an artist representation, in cartoon fashion, of various types of orbits. One should recognize that this graphic is not to scale, since the GEO satellite would be well off the page. This "compressed cartoon" is necessary since the GEO and even MEO satellites are, in fact, much, much further away and thus cannot easily be shown accurately in Fig. 2.4.

In sum there are a lot of different tradeoffs that can and indeed are made to determine what kind of satellite orbit is used to provide satellite communications services. These advantages and disadvantages are summarized in Table 2.1 below. One of the easiest choices is in the case of a domestic or regional satellite covering only a specific geographical area. Here a GEO satellite is clearly indicated.

The choice of a GEO satellite for domestic or even regional service is simply because a single satellite (plus a backup spare) can provide total coverage. Further, ground Earth stations can be pointed to the satellite and perform with high efficiency with no need for constant tracking systems, which increases cost and complexity as well as reliability issues. In short, LEO or MEO constellations, if they are indeed employed by a satellite operator, are designed and deployed for global type services.

GEO systems, which are by far the most numerous, can be used for domestic, regional or global systems alike. The following issues are typically considered in

Table 2.1 Advantages and disadvantages of various satellite orbits. (Chart supplied by author.)

Advantages of Low-Earth Orbit Systems
- Low latency or transmission delay
- Higher look angle (especially in high-latitude regions)
- Less path loss or beam spreading
- Easier to achieve high levels of frequency re-use
- Easier to operate to low-power/low-gain ground antennas

Disadvantages of Low-Earth Orbit Systems
- Larger number of satellites to build and operate
- Coverage of areas of minimal traffic (oceans, deserts, jungles, and polar caps
- Higher launch costs
- More complicated to deploy and operate – also more expensive TTC&M
- Much shorter in-orbit lifetime due to orbital degradation

Advantages of Medium-Earth Orbit Systems
- Less latency and delay than GEO (but greater than LEO)
- Improved look angle to ground receivers
- Improved opportunity for frequency re-use as compared to GEO (but less than LEO)
- Fewer satellites to deploy and operate and cheaper TTC&M systems than LEO (but more expensive than with GEO systems)
- Longer in-orbit lifetime than LEO systems
- Increased exposure to Van Allen Belt radiation

Disadvantages of Medium-Earth Orbit Systems
- More satellites to deploy than GEO
- More expensive launch costs than GEO
- Ground antennas are generally more expensive and complex than with true LEO systems
- Coverage of low traffic areas (i.e., oceans deserts, jungles, etc.)

choosing the right type of satellite orbit. As can be seen in Table 2.1 below, there are a number of factors to be considered such as launch costs, path loss, number of satellites and spares to be manufactured, operational and control complexities and perhaps most importantly the type of ground system to be utilized. (Note: Mobile systems are different from broadcast and fixed telecommunications satellite systems, because the antennas employed by users of the system are moving around rather than remaining stable and fixed at one point with an unobstructed and constantly clear view of the satellite above.)

The first choice in terms of choosing an ideal satellite orbit or constellation, however, may simply not be available. The lack of choice is constrained by orbital crowding. There are already over 300 communications satellites in operation, with most of these in the crowded GEO or Clarke orbit. Sometimes one must choose an orbit or constellation configuration that is not ideal. The decision thus becomes the challenge of finding a satellite location or constellation design for multiple satellites that can meet projected needs. The satellite system designer is charged with achieving a cost-effective design that provides the best solution after considering all the factors in Table 2.1 and more.

GEO, MEO and LEO are the prime configurations for satellite communications systems. There are hundreds of GEO satellites for global, regional, and domestic satellite communications systems as well as to serve military or defense communications purposes. There are also a growing number of MEO and LEO constellations. Although there are far fewer of these types of networks they require many more satellites to populate such a network and achieve global coverage. For instance the Globalstar mobile satellite system requires 48 operational LEO satellites plus spares, and the Iridium mobile satellite system requires 66 operational LEO satellites plus spares.

Beyond GEO, MEO and LEO systems, there are still some other orbits that have been proposed or actually used for special communications purposes. These are briefly listed and defined here.

Equatorial Circular Orbit (ECO)

This orbit, also known as the "string of pearls," is a circular orbit in the equatorial plane but deployed in MEO rather than GEO. Six to eight of these satellites with similar communications capability could continuously provide service to equatorial countries where some 2 billion people live. One satellite would continuously move into position to provide service as another satellite would rotate away from the current service area. This type of system has been proposed by the Brazilian space agency but not actually deployed. There have been proposals from Japan to create an extremely high-speed orbital network via a ring of satellites that are linked together via laser-based inter-satellite connections to achieve global interconnectivity.

Highly Elliptical Orbits (HEO), Extremely Elliptical Orbits (EEO), Molniya Orbits and Loopus Orbits

These are all very long elliptical orbits that have very long "hang times" above Earth's surface. These satellites can appear to be essentially in the same location for 8 to 12 hours at a time, especially at locations that are near the polar extremes. In short, this type of configuration can work particularly well for countries located at high latitudes such as Russia or New Zealand. Three satellites in this type of orbit can provide continuous world coverage.

The truth of the matter is that this type of orbital system works only for high-latitude countries because the satellites are very hard to track as they move back closer to the equator. When these satellites are at perigee they are zooming very fast indeed. The original Russian satellite system named "Molniya" used this type of orbit. The most exotic orbit of this kind is what is called the Loopus Orbit, which can serve different parts of the Northern Hemisphere with very long "hang times" to provide telecommunications or direct broadcast type services. There could also be an "inverse Loopus" geared to serve the southern latitudes. (See Fig. 2.5.)

Fig. 2.5 The Loopus-type orbit to serve northern latitudes. (Graphic supplied by author.)

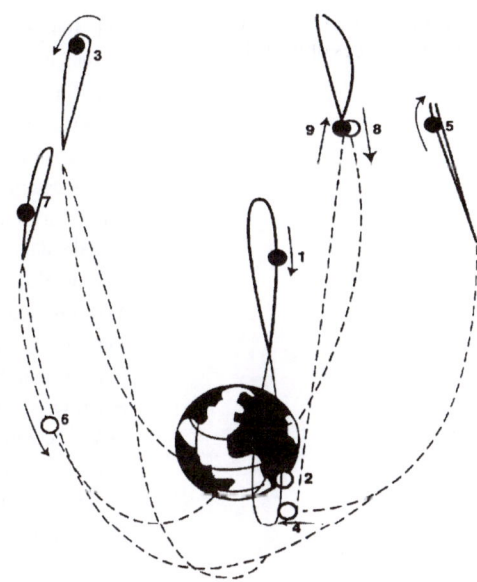

Quazi-Zenith or "Figure 8" Orbit

This is essentially a GEO that is inclined some 45 degrees. Three satellites in this orbit provide excellent high-look angles to countries located at 45 degrees latitude. This orbit is being used by Japan for mobile satellite communications and to provide supplemental space navigation services.

Super Synchronous Orbits

This is an orbit that is very difficult to track and would be used mostly for defense or military purposes.

Satellite Services and Applications

The types of services and applications that can be provided from an artificial satellite are continuously expanding. In terms of revenues, the most important communications satellite services are known by definitions provided by the International Telecommunication Union (ITU). These key ITU defined services are: Fixed Satellite Services (FSS), Broadcast Satellite Services (BSS) and Mobile Satellite Services (MSS). As one can see in Table 2.2 below there are also many other ITU-defined satellite services

Table 2.2 Officially defined
satellite services of the ITU

ITU Defined Satellite Services[4]
Fixed Satellite Services (FSS)
Inter-Satellite Services (ISS)
Broadcast Satellite Services (BSS)
Broadcast Satellite Services for Radio (BSSR)
Radio Determination Satellite Services (RDSS)
Radio Navigation Satellite Services (RNSS)
Mobile Satellite Services (MSS)
Aeronautical Mobile Satellite Services (AMSS)
Maritime Mobile Satellite Services (MMSS)
Maritime Radio Navigation Satellite Services (MRNSS)
Land Mobile Satellite Services (LMSS)
Space Operations Satellite Services (SOSS)
Space Research Satellite Services (SRSS)
Earth Exploration Satellite Services (EESS)
Amateur Satellite Services (ASS)
Radio Astronomy Satellite Services (RASS)
Standard Frequency Satellite Services (SFSS)
Time Signal Satellite Services (TSSS)

The focus of this book is on commercial satellite communications technology and services, and a review of Fig. 2.6 below reveals the broad range of commercial satellite applications now provided around the world. The satellite communications industry in 2010 achieved total annual revenues of some $170 billion (U. S. dollars). (This total includes military and defense satellite operations, plus all types of communications satellite services, satellite manufacture, launch services, Earth station manufacture, and launch insurance.) The communications satellite industry is clearly by far the largest. Fig. 2.6 clearly indicates that there are continually expanding satellite applications markets. Space applications today include space navigation, remote sensing, surveillance and meteorological observation. There is speculation that there will be, within the next decade, new offerings in solar power satellites as well as travel to in-orbit space habitats.

The following "official" definitions are provided for the "Big Three" in satellite communications services, which are Fixed Satellite Services (FSS), Broadcast Satellite Services (BSS), and Mobile Satellite Services (MSS).

The *Fixed Satellite Service (FSS)* is defined as follows: "A radio-communication service between Earth stations at given positions, when one or more satellites are used; the given position may be a specified fixed point or any fixed point within

---

[4]Joseph N. Pelton, *The Satellite Revolution: The Shift to Direct Consumer Access and Mass Markets*, (1998), International Engineering Consortium, Chicago Illinois.

Fig. 2.6 The broad range of expanding global satellite applications.[5] (Used with copyright permission by author.)

specified areas; the fixed-satellite service may also include feeder links for other space radio-communication services."[6]

The ITU definition for *Broadcast Satellite Service (BSS)* is as follows: "A radio-communication service in which signals transmitted or retransmitted by space stations are intended for direct reception by the general public. In the broadcasting-satellite service, the term "direct reception" shall encompass both individual reception and community reception."[7] Closely related to the BSS service is the more recently defined *Broadcast Satellite Service Radio (BSSR)*. This involves a new

[5] Op cit, Joseph N. Pelton and Scott Madry, p. 182.

[6] ITU Radio Regulations, Article 1, Definition of Radio Service, Section 1.21, http://www.ictregu lationtoolkit.org/en"practiceNote.aspx?id=2824

[7] ITU Radio Regulations, Article 1, Definition of Radio Service, Section 1.25, http://www.ictregu lationtoolkit.org/en"practiceNote.aspx?id=2824

Table 2.3 Annual services revenues for satellite communications and space applications. (Chart prepared by author. Figures derived from report by the Futron Corporation on behalf of the Satellite Industry Association.)

REVENUES FOR YEARS 2004-2009 IN SATELLITE SERVICES[8] (In billions of (U.S.) dollars per year)						
Type of Satellite Service	2004	2005	2006	2007	2008	2009
Direct to Consumer	$35.8 B	$41.3 B	$48.9 B	$57.9 B	$68.1 B	$75.3 B
DBS Television	$35.8 B	$40.2 B	$46.9 B	$55.4 B	$64.9 B	$71.8 B
DBS Radio	$0.3 B	$0.8 B	$1.6 B	$2.1 B	$2.5 B	$2.5 B
Broadband Internet	$0.2 B	$0.3 B	$0.3 B	$0.4 B	$0.8 B	$1.0 B
Fixed Satellite Services	$8.9 B	$9.3 B	$10.7 B	$12.2 B	$13.0 B	$14.4 B
Transponder Lease	$7.0 B	$7.3 B	$8.5 B	$9.5 B	$10.2 B	$11.0 B
Managed Networks	$1.9 B	$2.0 B	$2.2 B	$2.6 B	$2.8 B	$3.4 B
Mobile Services	$1.8 B	$1.7 B	$2.0 B	$2.1 B	$2.2 B	$2.2 B
Remote Sensing	$ 0.4 B	$ 0.6 B	$0.4 B	$0.4 B	$0. B	$1.0 B
TOTAL	$46.9 B	$62.8 B	$62.0 B	$72.6 B	$84.0 B	$93.0 B

radio frequency allocation that is used for audio broadcasting services. Today Sirius XM Radio (representing the merger of XM Radio and Sirius Radio) plus Worldspace both provide this type of satellite service. These systems provide audio-only broadcast satellite service either to automobile-based radios or handsets capable of direct reception from high-powered satellites. (This service is also called Satellite Digital Audio Radio Services (SDARS) by the U. S. regulatory authority, the Federal Communications Commission.)

The *Mobile Satellite Service (MSS)* is defined as follows: "A radio-communication service (a) between mobile Earth stations and one or more space stations, or between space stations used by this service; or (b) between mobile Earth stations by means of one or more space stations.This service may also include feeder links necessary for its operation."[9]

In general the trend in satellite service growth is quite positive. This is shown in the 2010 report from the Satellite Industry Association on the state of the industry as prepared by the Futron Corporation. (See Table 2.3 below.) Although there is upward growth in all the markets, what is most striking is the predominance of the direct-to-consumer services that are provided to end users. The size of revenues in what might called satellite "retail sales" that provide services directly to consumers

---

[8] Satellite Industry Association "Executive Summary, 2010 State of the Satellite Industry Report", Prepared by the Futron Corporation, 2010 Washington, D. C. www.sia.org/news_events/pressrele ase/2010StateofSatelliteIndustryReport2010(Final).pdf NOTE: There are many reference books and almanacs that provide a guide to satellite communications systems in operation around the globe. See, for example, the appendices in Handbook of Satellite Applications by Joseph N. Pelton, Scott Madry and Sergio Camacho Lara, 2012, published by Springer in New York.

[9] ITU Radio Regulations, Article 1, Definition of Radio Service, Section 1.XX, http://www. ictregulationtoolkit.org/en"practiceNote.aspx?id=2824

totally dominate and outweigh what might be called the satellite "wholesale markets" that provide services to large telecommunications organizations that in turn resell these services to end users.

The global report prepared for the Satellite Industry Association by the Futron Corporation is the most thorough and complete report available. This report is unique in that it builds up its data from individual annual revenue reports from over seventy commercial operators.

Nevertheless, the reported revenues from this report still require some interpretation. First of all, revenues related to space navigation markets are not included. It is estimated by some that this market, in terms of direct and indirect sales of space navigation devices plus applications in map making, transport systems, law enforcement, defense and security operations, etc., could be as large as $5 or $6 billion per year.

The above figures from Futron also do not include expenditures and costs related to national space systems operated by governments nor do they include expenditures related to military or defense satellite systems operations, except in the case of "dual use" of commercial systems. Also, it should be recognized that the reported revenues differ between one satellite service and another.

The total revenues related to mobile satellite services, for instance, are more dependent on the sale of satellite handsets, consumer mobile user units, etc., than to the other satellite services in terms of "sizing the market." User transceivers tend to be updated more frequently and thus turn over more quickly in the mobile satellite industry than in the other space communication services, and the total number of units is quite large in comparison to, say, fixed satellite services.

Further, the mobile satellite industry is in the midst of a major change with the launch of new hybrid mobile satellite networks that will include the so-called "ancillary terrestrial component." This new aspect of the mobile industry will likely increase total revenues in this sector. Nevertheless, with the active combining of "terrestrial wireless and satellite" service it will be much more difficult to determine which revenues relate to space- and which relate to ground-based service. This is all to say that the mobile satellite revenues, as reported by the Satellite Industry Association, may currently give a somewhat minimized view of the total size of this part of the satellite world and that revenues in this sector can be expected to increase the most rapidly in the next five years if the new hybrid systems prove successful.

Overall, Table 2.3 above demonstrates not only the large size of the various satellite communications service sub-markets but also just how dominant this space industry is in relation to other emerging commercial space markets such as remote sensing. Today, space navigation services probably represent at least $5 billion (in U. S. dollars) in total worldwide sales of services and equipment. These services are perhaps the most rapid growing new space application market, but clearly satellite communications services are by far the largest single commercial space market. Further, it should be noted that satellite communications services are only a part of the total market. When ground systems and user equipment manufacturing, satellite spacecraft manufacturing, launch services, launch insurance and other support industries are taken into account, the satellite communications industry annual sales swell to some $170 billion (U. S, dollars) in total revenues, as will be discussed further in Chapter 6.

Overview of Communications Satellite Technology and Operation

3

Introduction

The original idea of a communications satellite that could be deployed in geosynchronous orbit was a quite simple one. The concept was to create the equivalent of a very, very tall microwave relay tower in the sky. This virtual tower could connect voice, messages or data links across oceans or could be a broadcast station for radio and television with very broad coverage. As noted in Chapter 2, a communications satellite in GEO remains virtually motionless with respect to its sub-satellite point and thus is perfectly positioned to capture uplinked signals with its receiving antenna(s) and translate these into downlink radio signals to be sent through its transmitting antenna(s). The same frequencies cannot be used for the uplink from Earth and the downlink return because they would interfere with one another. These are thus separately allocated for the up- and downlink channels.

This chapter provides basic technical information about communications satellite engineering. First it addresses how various types of communications satellites operate and the importance of designing a communications subsystem with an adequate link budget and associated power margin to assure continuity of services. Secondly it explains technology that is essential to how a satellite carries out its mission. This explanation will include the tracking, telemetry, command and monitoring (TTC&M) systems that keep the spacecraft functioning properly. Finally there will be a "soup to nuts" presentation of the complete lifetime of a satellite from conception, design and manufacture through to deployment, operation and end of life activities.

The Communications Subsystem

The heart of a communications satellite is its communications subsystem. This includes both the receiving and the transmitting antennas on the communications satellite plus the associated electronics that allow the satellite to perform its basic function. After the signal is received by a communications satellite it is "filtered" or

J.N. Pelton, *Satellite Communications*, SpringerBriefs in Space Development,
DOI 10.1007/978-1-4614-1994-5_3, © Joseph N. Pelton 2012

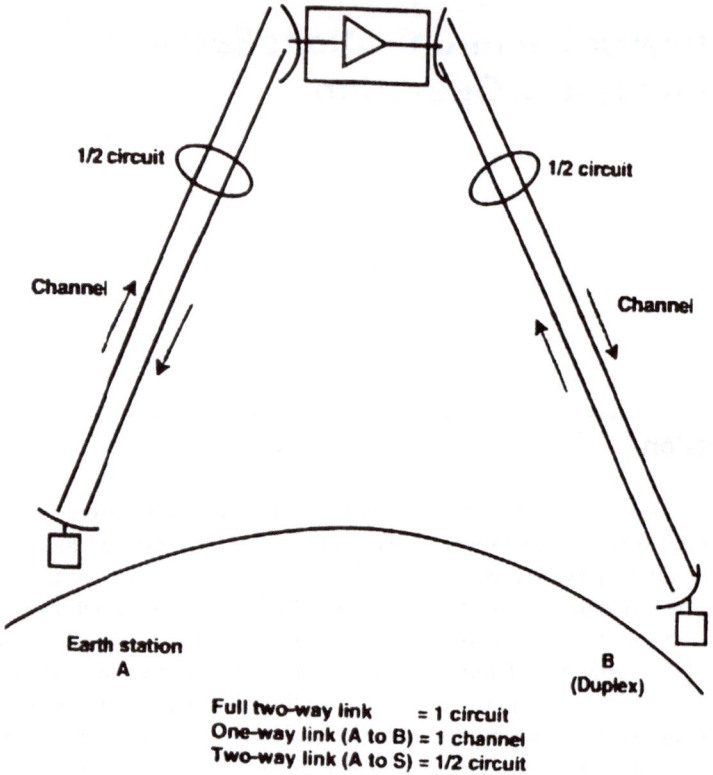

Fig. 3.1 Uplinks and downlinks for a GEO communications satellite. (Graphic by Joseph Pelton.)

"processed" to reduce signal distortions. The signal is then translated to the downlink frequency and amplified. Some form of amplification usually occurs both in the processing of the uplinked signal and again with the downlink signal. The downlink signal is then channeled into a feed system. In most satellites today the feed system has many feeds that allow many different beams to be formed off of a multi-beam satellite antenna. Thus the downlink signal must be routed to the correct feed so that the downlinked signal can be transmitted back to Earth within the correct antenna beam to reach the correct destination. (See Fig. 3.1 below.)

These satellite beams might be a "global beam" that covers from GEO about a third of Earth's surface, a "zonal beam" that may cover a continent or a part of a continent, or a spot beam that is targeted to a part of a country, a part of the ocean (for mobile communications satellites) or even a metropolitan area.[1]

---

[1] Gary D. Gordon and Walter Morgan, *Principles of Communications Satellites*, (1991), John Wiley and Sons, New York p128

As noted earlier, there are GEO systems that generally support direct broadcast, fixed satellite (i.e., the ground antennas are fixed in place and do not move) and mobile satellite services. There are about 300 active GEO satellites in orbit today and these are carrying out a wide range of satellite communications functions for commercial civilian or defense-related purposes. There are also quite a number of MEO or LEO constellations. For the most part these MEO or LEO communications satellite constellations are essentially designed for carrying out mobile satellite services or to support machine-to-machine (M2M) services. In the case of an LEO or MEO constellation the satellite antennas are constantly moving with respect to any particular point on Earth, and the antenna for the mobile user could be located on an aircraft, a car, truck, train, ship or even carried by a pedestrian. In all cases the antenna is moving relative to the satellite.

As we noted in the previous chapter, the advantage of an LEO or MEO constellation is that the free path loss due to the spreading of the downlink beam from the satellite antenna to the ground antenna or the uplink beam from the ground antenna to the satellite is much less than in the case of an GEO satellite. This means that successfully closing the communications link between the satellite and the Earth station is much easier to achieve in a MEO and especially at LEO constellation. Also as we noted in Table 2.1 there is also much less transmission delay or latency.[2]

This is the good news. There are, however, also major disadvantages in terms of the design and performance of the user antennas for mobile satellite systems using either an LEO or MEO constellation design. The user antennas have to be able to capture signals as the satellite moves across the sky. In the case of a GEO satellite the available beam is constantly fixed and a mobile user is not constantly switching from beam to beam since the user largely moves within the same beam. Regardless of whether it is a GEO, MEO or LEO system, the user antenna for mobile satellite services must either "track the satellite" or the user antenna must be designed to be nearly an "omni" type antenna that can receive a satellite signal from all angles across the sky, regardless of where a "visible" satellite moves.

This means that the "omni-type" ground antenna for mobile satellite services (unlike the case for FSS and BSS services) cannot be high gain. In fact if the user antenna cannot track, it is quite low in gain. Clearly one option is for the mobile satellite user antennas to be able to rapidly track the satellite as it moves across the sky and to have an antenna mounted on an ship, aircraft or vehicle that can "follow" the satellite's trajectory. For small hand-held devices this is, of course, really not possible. For hand-held transceivers the antenna for transmission and reception is essentially an "omni-antenna" that is designed to receive a signal from any angle, or at least any direction that is above the horizon. In order to "close the link" between the satellite and a hand-held unit operated by a consumer, the transmitting power from the satellite has to be quite high in order to compensate for the modest capability of the user device.

There is another problem with the LEO or MEO constellation, that the satellite is only visible for a brief period of time until it must be replaced by another satellite.

---

[2] Mark Williamson, *The Communications Satellite*, (1990) Adam-Hilger, London, U.K. pp. 354–355.

In the case of an LEO constellation this visibility time is only a few minutes, and in the case of spot beam operation, the switchover from one beam to another can be as brief as only one minute. The ability to switch from beam to beam and then from satellite to satellite is a challenge particularly if at the exact time of switchover the user might be going into a tunnel or driving into a heavily wooded area.

These complications have driven some satellite designers to develop very large GEO satellites with huge high-gain antennas to support communications to handheld units and without the need to switch frequently between beams. The bottom line here is that the most difficult system to design is a mobile communications satellite system. The challenge is to design a satellite system that allows the consumer to have a transceiver that is virtually as small as a cell phone and provide reliable communications. Fixed satellite systems (FSS), broadcast satellite systems (BSS) and store and forward machine-to-machine (M2M) data relay satellites are much easier to design. For fixed satellite systems in GEO one can line up an Earth station with a satellite that is constantly visible and there are no obstructions between the satellite and ground antenna facility. If you know the performance capability of the antenna on the ground and on the satellite then one can calculate the transmitting power to complete this link. This is called the power link margin. There are lots of computer programs available today where you can plug in the key information and it will tell what the power link budget will be. You can then add some power link margin against bad weather when reception and transmission is more difficult and you are all set. The world of mobile satellites is more difficult because one must deal with a lot more variables.

It is clear then that the easiest task is to calculate the link budget and to determine what would be an adequate "power link margin" for a GEO satellite. In this case if you double or perhaps quadruple the amount of power that allows you to complete a satellite link then you can have a reasonably good expectation that you will have high link availability. This is the case when the antenna is fixed and constantly pointed and there is not a problem of a constantly changing environment. You do not have to worry about trees, skyscrapers, billboards, utility poles or mountains getting in the way.

In all forms of satellite communications there is a provision for "power link margin," that provides a degree of additional power beyond the link budget that completes the link between the satellite and the user terminal. The obvious purpose of having a power link margin is to provide "insurance" to maintain something like over 99% link availability. A power link margin such as 3 dB provides twice the power needed to close the link or 6 dB provides four times the power needed to close the link. This is the normal type of link margin that one finds with GEO services to support fixed or broadcast satellite service.

This type of margin (i.e., 3 dB to 6 dB) provides spare transmission power to complete the link that compensates for any problem such as rain attenuation or poor atmospheric conditions. The setting of power link margin for FSS or BSS services is a much simpler proposition than is the case of mobile satellite services. In the case of MSS systems, especially with LEO or MEO satellite constellations, the satellites are constantly moving closer or further away. But the challenge gets even more difficult. The user's terminal is also constantly moving, and the transceiver's antenna may at any moment be going by a building, a billboard, into a tunnel or a dense forest. The most difficult challenge of all is when the user enters a concrete and steel building and wishes to talk from that location. One solution that is just beginning to be implemented is to design a hybrid terrestrial and satellite network so that high-powered terrestrial cell towers provide service within a city, and then in suburban and rural areas the mobile satellites can provide the service when terrestrial cell towers are not available. This is called mobile satellites with ancillary terrestrial component, at least in the United States, where these systems are first being implemented. It is called Complementary Ground Component (CGC) in Europe, or it may be simply called a hybrid satellite and terrestrial cellular mobile communications service.

The question of great interest to any designer of a satellite system is what level of quality performance is to be sought. In the modern world of digital communications the "quality performance standard" is now usually defined as a bit error rate. This is the ratio of digital bit errors in comparison to the correct digital bit transmissions. Usually the ratio sought in modern communications is no more than one error per million bits transmitted as defined by the Integrated Services Digital Network (ISDN) standards.

The other key standard is for system availability. This is simply defined as the percentage of time the service is available as compared to the time when it is not available. The key to answering the question as to what service standard can be provided by a satellite system is undertaken first by calculating a power link budget. In practice this is usually just called a "link budget." Once this calculation is undertaken and the result is known, the additional question is to establish what additional power margin should be reasonably provided to keep the signal intact even if there are rainstorms or atmospheric conditions that might interfere with the signal getting through.

So how does one calculate a link budget? Most satellite engineers would simply use a set of specialized software. This consists of a spreadsheet with all of the factors to be considered and one simply plugs in the appropriate data. The single biggest factor is, of course, the "path loss" that occurs as a satellite signal travels the very long distance between the satellite and the user terminal – especially in the case of a GEO satellite. The minimum distance a signal travels between a GEO satellite and a ground antenna is 35,870 km. This is, in fact, a rare condition because the ground station is very rarely at the sub-satellite point. Even ground stations that are located on the equator are very rarely exactly underneath a GEO satellite since the ground station is typically located either east or west of the orbiting spacecraft. In calculating power link budgets one must thus also consider the case where the ground antennas are located well away from a point that is exactly right below the spacecraft – called the sub-satellite point. In fact the worst case, in terms of performance is when the transmitting and receiving earth station are the maximum distance away from the satellite, but both ground stations can still "see" the satellite. One might consider the most extreme case where the up-linking Earth station is 60 degree away from the sub-satellite point in longitude as well as 60 degrees away in latitude. Further let's consider that the receiving Earth station is also the same distance away, perhaps in the opposite direction. In such a case the total distance from the Earth station to the satellite and then back to the other Earth station could be well over 100,000 km.

When satellite signals travel these huge distances the original beam spreads out and the loss of signal strength due to path loss is indeed huge. The signal is

Table 3.1 The ten elements included in a link budget calculation. (Chart prepared by Joseph Pelton.)

The Ten Factors Used In Calculating a Communications Satellite Power Link Budget
• Transmitting power at the satellite antenna
• Satellite antenna gain relative to an isotropic radiator
• Equivalent isotropically radiated power (known as E.I.R.P.)
• Illumination level at the receiver
• Free space path loss (equivalent to the square of the longest transmission path)
• System noise temperature
• Figure of merit for receiving system
• Carrier-to-thermal noise (C/T)
• Carrier-to-noise density ratio
• Carrier-to-noise ratio (C/N)

diminished many billions of times. Also the delay in the transmission becomes not a quarter of a second but as much as one third of a second.

It is because of this huge loss of signal strength that communications engineers find it convenient to calculate in decibels. The decibel is an exponential system that allows calculations to cope with differentials that can be billions of times higher or lower with much greater ease. For example a loss of 10 dB is equivalent to a loss of 10. A loss of 20 dB is a loss of 100. A loss of 30 dB is a 1,000. A loss of 80 dB is loss of 100 million. A loss of 100 dB is a loss of 10 billion.

There are ten key factors that go into link budget calculations as identified above in Table 3.1, where specific information can be obtained and put into a link budget calculation. These factors include values such as transmit power, the gain of the satellite antenna, the gain of the ground antenna, and, of course, the largest single factor, namely the path loss between the satellite and the ground antenna or vice versa. The use of specialized spreadsheet software helps to ensure that no single factor is overlooked and also aids in avoiding mathematical errors.

The main thing to keep in mind is that it is always the worst case condition that one must consider and not the best case conditions when considering a link budget as well as subsequently considering an appropriate link margin. Thus key questions that one must ask include the following: What are the extremes of the transmission paths between the transmit antenna and the receive antenna in terms of distance? Of all the antennas that will access the satellite system, what will be the lowest gain user-antenna that must be taken into account?[3]

In a way the link budget calculation is the easy part when one is designing a satellite. Today most engineers plug in all the figures and the computer spits out the calculated answer. The hard part is in considering what is a reasonable "link power margin" when one engineers the total system performance.

The consideration of the power link margin must take into account those factors that vary with the environment and are not fixed amounts. Heat scintillations, rain

---

[3] *Op cit*, Gary D. Gordon and Walter Morgan, p. 34.

fade, and other atmospheric conditions can greatly reduce performance. Generally, the further one goes up to higher frequencies and smaller wavelengths the greater this problem becomes and the higher the link power margins that will be required. In the case of mobile satellite systems in LEO or MEO constellations or even GEO systems there is no guarantee that the user or the car carrying the person using the mobile satellite transceiver will not be going underneath trees, going into a building or even entering a tunnel. The addition of "power link margin" as one goes to higher frequency is certainly a calculation that can be made on probabilities. For satellite communications at the 6 and 4 GHz range power margin is often double or 3dB. For satellite communications at the 14 and 12 GHz range power margin is often quadrupled to 6 dB. For satellite communications at the frequencies above 28 GHz a link power margin of 9 dB (8 times) or more is used.

The issue of link power margins for mobile satellites is much more complex and difficult. In this case the transceiver is design to "hold the call" while there is a drop in power for quite a few seconds. This assumes that one might clear an obstacle after 10 or 20 seconds. The other precaution is have a very high link power margin of maybe 10 to 12 db (i.e., 10 to 16 times the power needed to complete the link). This high power margin assumes that indirect signals may bounce off of other surfaces and reach the mobile satellite communications device. In the case of terrestrial wireless systems, however, it is now possible to design systems that have 30 dB power margin (i.e.,1,000 times the needed power to complete the link). Those who are used to this much power link margin can have an unrealistic expectation as to what a satellite phone can achieve in terms of reliable link performance. It is this issue of power link margin that has pushed the mobile communications industry toward hybrid systems that rely on terrestrial wireless networks within the city with large link margins and then to use satellites in rural and remote areas.

The problem is that there is not a clear "right answer." For instance a GEO system that is operating with a 6 GHz (uplink) and a 4 GHz (downlink) can probably maintain reliable service with only a 3dB margin. This is to say that the satellite will have twice the performance capability as needed to complete a link. Even if there is a low elevation angle up to the satellite and a low elevation angle back to the receiving Earth station this is probably adequate to insure reliable service. This 3dB margin would probably be considered reasonable because at C-band (i.e. 6GHz and 4 GHz) there is not a much precipitation attenuation to degrade the signal strength. Further the antenna coverage of the satellite beams tend to be quite broad and thus transmit power falls of only gradually from the center of the beam so that users at the edge of the beam would still be only experiencing a 1 or 2 dB degradation.

Another engineer might be designing a communications satellite to operate at 30 GHz uplink and a 20 GHz downlink. In this case precipitation attenuation at these very high radio frequencies represents a very significant problem because so-called "rain fade" becomes a greater problem at higher frequencies. This problem, of course, becomes even worse in a very rainy climate with yearly monsoons. Also the spot beams would likely be much smaller with a more rapid fall off of power at beam edge. For this type of satellite design a 6 dB or even 9 dB margin (i.e., four to eight times the power to complete the link) might be chosen. In fact, a more "intelligent" link margin might be developed. In this case there might be power margin available on-board the satellite that could be allocated "on demand" to particular spot beams. This would be a "responsive link margin on demand." In this type of satellite design the additional power margin could be delivered to particular spot beams where there is heavy rainfall and rain attenuation is a particular problem. For example, there might be 3 dB of generalized power margin and another 6 dB power margin that could be provided to selected beams where a problem with precipitation attenuation or atmospheric conditions warranted the allocation of more power.

The most difficult case of all would be that of the mobile satellite system. This is the case where there are a very large number of variables to cope with indeed. This large number of variables involved with mobile satellite systems makes the choice of an optimum power margin very difficult. In this case one might find power margins that run from 10 to 20 dB depending on the actual type of service that is to be offered in terms of availability, bit error rate and whether one intends for callers to be able to complete and hold a connection from inside of a building or from inside of a vehicle. One could, of course, always opt for the largest possible power margin, but this involves a very large cost. A satellite system with twice the power link margin added to it may not cost twice as much as the projected cost of building the system with a lower link margin, but there is clearly a cost penalty to be paid. The doubling of a power margin might add a third or even half as much to the overall cost when one adds up the additional engineering, manufacturing and launch costs.

Today most "C band" satellite systems in GEO provide for a 3 dB link margin. GEO systems that must contend with significant rain attenuation problems in the Ku and Ka bands tend to provide for a 6 dB to 9 dB link margin. Systems for mobile satellite communications, regardless of whether they are LEO, MEO or GEO, try to provide at least 10 to 12 dB of link margin and sometimes considerably higher. The problem is that terrestrial cellular mobile communications systems now provide up to 30 dB of margin in urban areas where there are steel and concrete skyscrapers. Mobile communications satellite systems thus need to provide ever higher link margins to be competitive. The other option is to design hybrid systems so that terrestrial networks are deployed inside cities while satellites with broader coverage and somewhat lower link margins are available in suburban and rural areas.

The truth of the matter is that this is a very complicated issue. For instance, systems that provide global mobile satellite services have to worry not only about providing sufficient power to meet consumer needs but also they must also limit their interference with satellite and terrestrial communications systems as well as other users of closely adjacent frequencies such as space navigation systems and radio astronomy. This is to say that even if they wished to provide additional power link margin, they might be operational or technically constrained by intersystem coordination requirements. If the satellite irradiates too much power it could cause harmful interference to other satellites, to terrestrial communications systems or very sensitive instruments such as radio telescopes.[4]

---

[4] Dennis S. Roddy, *Satellite Communications*, 3rd Edition (2001) McGraw Hill, New York. Pp. 307–335.

Spacecraft and Major Subsystems

The design of a communications spacecraft is a complicated affair. All of the various parts of the spacecraft support each other. These many subsystems are interdependent in sophisticated ways. If the right components are not available with the right functionality and scale the satellite can fail and may never recover. Critical parts include the power supply and power converters, the thermal controls, the surge protectors, wiring and switches, the sensors for determining orientation and positioning, the stabilization devices, and backup electronics. Even the spacecraft structure and physical deployment devices for the antennas and the solar arrays are critical. This is not to overlook the tracking, telemetry and command and monitoring (TTC&M) systems, but these will be discussed separately since they involve a system that is partially on board the spacecraft and partially on the ground at terrestrial antennas and control centers.

The Spacecraft Structure and Physical Deployment Devices

The first satellites were simple devices that looked like high-tech beach balls or oil cans that had come out of a laboratory. These devices, such as *Sputnik, Telstar* and *Relay* were essentially experiments to prove that satellites could work. They had little surface on which to house solar cells and the simplest of antennas to deploy.

When operational communications satellites were designed and deployed, the spacecraft structure was constantly expanded to support more and more solar cells, to host larger batteries and more sophisticated electronics and to support the deployment of higher gain antennas that could be constantly and accurately pointed toward Earth. This led first to the "de-spun satellite" design. With this design the interior of the satellite that supported the high gain antenna system could spin in one direction at 60 rpm while the outside body, coated with solar cells, could spin in the opposite direction at 60 rpm in order to keep the spinning platforms upright and the antenna system, in effect, constantly pointed accurately back to the targeted areas below. This design in turn gave way to 3-axis body stabilized spacecraft structures. These spacecraft with a boxlike structure were maintained in place with very rapidly spinning momentum wheels or inertial wheels (at high speeds of 4,000 to 5,000 rpm). From this box, which contained all of the key subsystems, it was possible to extend either the solar arrays or various antenna systems with high efficiency. The solar arrays were extended as if they were wings protruding outward from the side of the box, while the antenna systems were deployed from the "top" or "bottom" of the box. (In space such terms really have no meaning.)[5]

Figs. 3.2, 3.3 and 3.4 show the progression in ever more efficient ways to deploy solar cells to achieve the maximum amount of solar irradiation. First, the solar cells

---

[5] Ibid. pp. 173–177.

Fig. 3.2 The small beach ball-like *Telstar* satellite with solar cells mounted on the outside of the spacecraft, as shown prior to launch in 1962. (Photo courtesy of NASA.)

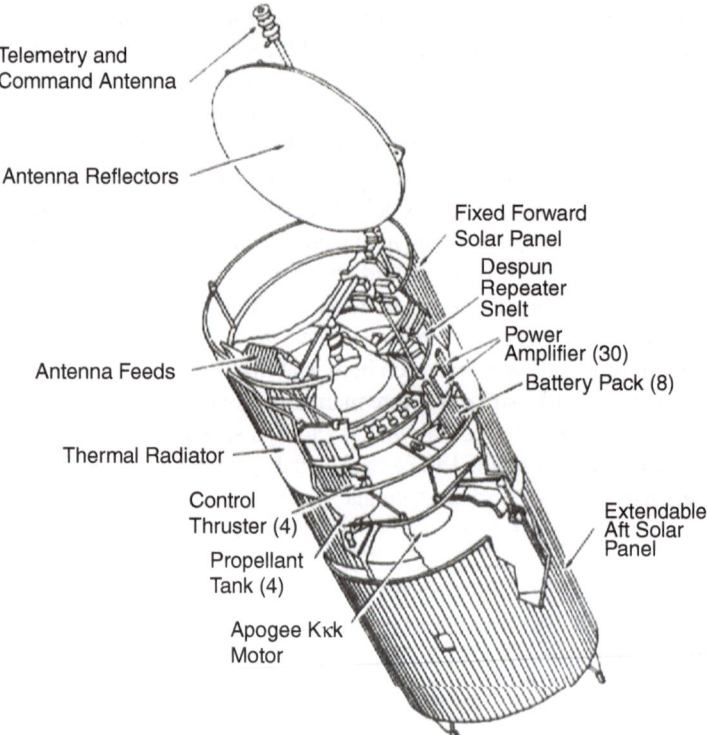

Telemetry and
Command Antenna

Antenna Reflectors

Fixed Forward
Solar Panel

Despun
Repeater
Snelt

Power
Amplifier (30)

Battery Pack (8)

Antenna Feeds

Thermal Radiator

Control
Thruster (4)

Extendable
Aft Solar
Panel

Propellant
Tank (4)

Apogee Kκk
Motor

Fig. 3.3 A large scale de-spun satellite of the early 1990s. (Graphic courtesy of *Intelsat*.)

Fig. 3.4 An advanced
three-axes body stabilized
space navigation (NAVSTAR)
satellite with deployable solar
arrays

were mounted directly on the satellite (i.e., the *Telstar* satellite). Next solar cells were mounted on the outside drum of "spinners." In such a design a drum skirt could be dropped down to increase the size of the solar array (i.e., *Intelsat 6*). Finally there are spacecraft that have extended solar arrays that project from a three-axis body-stabilized spacecraft. In this type of design the solar array could be extended perpendicularly to the satellite body and thereby be deployed at different angles to get maximum exposure on a constant basis except when the satellite was in seasonal eclipse.

Soon satellites became much more sophisticated. They were designed to be stabilized on three axes so that the satellite platform could accurately and continuously be pointed so high that gain antennas could illuminate precise locations on Earth's surface.

It became clear in time that a three-axis body stabilized design that allowed continuous illumination of solar arrays by the Sun and more precise pointing was the best design.

The Power Subsystem

Today's communications satellites typically are powered with large-scale solar arrays that are deployed after the satellite reaches its operational orbit. There is also backup power provided by batteries during eclipses or momentary lapses in the solar array power supply. Significant progress has been made to develop batteries that have greater power density storage capacity and longer life. Today the best systems are rechargeable lithium ion batteries that can maintain lifetimes of over 15 years with high-power charge densities. Progress is being made to develop fuel cells that may be able to outperform batteries, but their costs and lifetimes will need to be improved to replace batteries.

Improved solar array systems are also being developed. Photovoltaic cells using materials such as gallium arsenide are now being deployed. Also solar cells with an increased number of photovoltaic junctions are being manufactured. These cells can convert a wider spectrum of power from sunlight, in particular being able to convert the most energetic photons in the ultraviolet part of the spectra into electricity.

Also there are improved "roll out" solar arrays that are more reliable and lighter in weight than conventional solar array deployment systems. It is now possible to design and deploy solar array systems with power capabilities as high as 12 to 20 kilowatts.

It is thought by some, however, that eventually it will be necessary to convert from solar arrays to higher powered energy sources as satellites continue to increase in size and power requirements. Nuclear and isotopic power systems are, of course, quite controversial, but research continues to explore nuclear propulsion as well as nuclear power sources for the largest spacecraft. For the foreseeable future solar arrays, with batteries providing backup power during eclipses, will continue as the prime power sources for communications satellites.[6]

The Thermal and Space Environmental Control Systems

The hostile environment of outer space represents a true challenge to spacecraft engineers. The heat and cold, cosmic radiation plus radiation within the Van Allen Belts, solar flares, and electromagnetic phenomena are just some of the environmental challenges that satellite designers face. The challenge is further increased by the desire to create satellites that can have a useful lifetime of 15 years or even longer. This is a quite difficult challenge in that repair and refurbishment capabilities are currently not available to satellites in Earth orbit.

One of the first challenges is to address the thermal environment. Outer space is very cold, and the temperatures encountered in outer space can approach absolute zero. On the other hand, the Sun's energy is still quite strong, even in Earth orbit. If a satellite does not have reflective surfaces a spacecraft can heat up to excessively high temperatures. Also the electronics inside the spacecraft body that are powered by solar arrays and batteries can heat up to excessively high temperatures if there are not means to channel heat out into space. The satellite must be designed to include reflective materials that help keep the spacecraft's temperature in reasonable balance so that the electronics, processing devices, and sensitive equipment within the spacecraft do not either overheat or grow too cool. The spacecraft thus typically includes "heat pipes" that allows heat from the inside of the spacecraft to transfer to the spacecraft edges and vent into outer space. The telemetry system for the satellite is connected to sensors that allow the "reporting" to satellite control systems of the internal spacecraft temperatures. If the temperature should rise or fall above set limits, alarms notify satellite engineers or expert system software of the problem and corrective actions are taken by reorienting the spacecraft to increase or decrease reflectivity or take other corrective action.

---

[6] *Op Cit*, Mark Williamson, pp. 97–115.

The Positioning and Orientation Systems

As noted earlier, the key to increasing satellite communications performance is the ability to orient the communications antennas so that they point precisely toward Earth. This is particularly critical for geosynchronous communications satellites, since they are so distant from Earth's surface, and thus transmission path losses are particularly great. This difficulty is overcome by having particularly high gain communications antennas that can direct the signal of these large aperture antennas precisely back to Earth's surface at carefully configured locations.

All of this complex design for geosynchronous satellites would not work if the satellite was not continuously and accurately pointed. A constant readout of telemetry and tracking information is undertaken to ensure that the various satellite spot beams reach their exacting and pre-set locations. Even MEO and LEO satellites are designed to send their beams Earthwards with a high degree of accuracy as the various satellites follow their constellation orbits. In the case of MEO and LEO constellations the same precision is not required because the satellites are closer to Earth and because the geometry of the beam patterns in global constellations are more forgiving of beam pointing error. Nevertheless the orientation, positioning and pointing requirements are still demanding for all types of communications satellites.

Geosynchronous satellites are typically designed today to point with an accuracy of a small fraction of a degree at an orbital distance of 35,870 km. Since these satellites are by far the farthest away the pointing accuracy for these satellites are the most demanding. A combination of Earth sensors, star sensors and RF beams helps to maintain the correct alignment, and rocket thrusters on the spacecraft can be fired to restore the correct pointing accuracy. MEO and LEO satellites are also oriented and pointed with some careful accuracy, but since these satellites are much closer the pointing accuracy is not nearly as crucial. An MEO or LEO satellite whose antennas were sending signals out into space and away from Earth would certainly be worthless. Pointing toward Earth, even for this type of satellite, is still quite important. The more beams transmitted by an LEO or MEO satellite (such as the 48 beams that come inward to Earth from an LEO-based Iridium satellite) the more critical the beam pointing accuracy on these spacecraft. If a global beam antenna that illuminates almost a third of the planet shifts a bit in its orientation it really does not have a great impact on the power received over a very wide area. On the other hand if there is a spot beam that is targeted to cover an area like greater Chicago a slight shift in the antennas pointing accuracy could create a loss in service.

The first challenge is to get the satellite positioned correctly in its desired location in orbit. If there is insufficient fuel to keep a satellite at the right east-west location in the GEO plane, or a lack of capability to keep it from moving off of the orbital plane in a north-south direction, then a communications satellite can become essentially worthless. The W3B communications satellite, successfully launched by the Eutelsat organization in November 2010 on an Ariane 5 vehicle, ended up as total failure. This satellite, valued at over $200 million dollars and scheduled to replace three satellites in the Eutelsat fleet, was declared a total loss as of November 7, 2010, because the fuel tanks for the spacecraft thrusters had leaked all of their fuel. Fortunately for Eutelsat it had obtained full insurance coverage, and the W3C satellite is scheduled to be launched as a near replacement satellite.

The type of thruster systems used to support the initial positioning, orbital station-keeping and removal from orbit have evolved in recent years. For many years spacecraft thrusters that were chemically powered have predominated. The most common propellant for spacecraft thrusters has been the quite noxious and poisonous and very explosive fuel known as hydrazine. Such thrusters systems are called monopropellant thrusters since their mini rockets used hydrazine fuel only. The thrust is achieved by the rapid expansion of the hydrazine as it is released from the thrusters' nozzle. In some hydrazine thruster systems there are heaters that maximize the hydrazine's super rapid expansion at its release, and this maximizes the impulse achieved by the hydrazine gas release when the thruster's valves open.

Larger satellites have for some time used dual-propellant systems that combine hypergolic fuels to create a strong chemical explosive force. These types of hypergolic propellants ignite when combined from separate fuel tanks with an oxidizer. These are thus known as bi-propellant systems, since in this case hydrazine is often combined with another explosive oxidizer. This chemical explosion produces a sizable thrust, but it also consumes a quite measureable amount of fuel as well as the igniting oxidizer. A large satellite will thus require hundreds of kilograms of the bi-propellant fuel to keep the satellite in the correction position and orientation – especially if there is a need to reposition a satellite from one operating position to another in rapid fashion. In recent years electric ion thrusters that support longer life, by assisting with station-keeping of the spacecraft, have become more and more common. These thrusters produce a very low force as ions are expelled from the spacecraft when the fuel is electrically heated to very high temperature. Rather than being a "chemical combustion" with lots of immediate "oomph," electric ion thrusters produce a very low propulsive force – but they do so for a much longer period of time. In the long run they provide more thrust per unit of fuel, but they have limited ability add immediate power.[7]

Satellite Sensors That Allow Accurate Positioning and Orientation

The sensors that assist with the accurate pointing of the high-gain communications antennas are a critical part of the spacecraft design. The various Sun, Earth and star sensors have become more and more precise and thus better able to stabilize spacecraft and their pointing accuracy. These sensors are important not only to maintain accurate pointing but also for recovery in the event an anomaly causes the satellite to lose its three-axis pointing direction.

The problem is that as high gain antennas have grown in aperture size and performance the need for ever better pointing accuracy has also grown. This has led to

---

[7] Joseph N. Pelton, *Satellite Communications 2001: The Transition to Mass Consumer Markets, Technologies, and Systems*, (2001) International Engineering Consortium, pp. 237–239.

the design of RF alignment and pointing systems to allow pointing accuracy in the range of 0.25 to 0.50 degrees and even better. These precise RF alignment systems allow even huge 18- to 20-meter communications antennas deployed on mobile satellite satellites to be oriented with pinpoint accuracy toward locations on Earth. The various sensors to detect stars, the Sun and Earth, however, remain important to help to reorient the spacecraft in the case it should lose the desired three-axis pointing required to maintain optimum performance.[8]

The Problem of Space Debris

The increasing amount of space debris is a threat to the future deployment and operation of communications and other types of application satellites. Today there are on the order of 30,000 pieces of space debris of 10 cm or more in diameter that are being tracked in polar and LEO, MEO or GEO. In fact, there are literally millions of pieces of debris in orbit that are on the order of millimeters in size – items the size of chips of paint. These microscopic chips traveling at thousands of kilometers an hour have sufficient destructive momentum to create a dangerous chink in the window of a space shuttle or even kill an astronaut by penetrating a space suit.

The problem of orbital debris has grown steadily in recent years, especially in the low Earth and polar orbits. Fig. 3.5 below represents a current graphic showing the ever-increasing build up of space debris in LEO.

Concerns about orbital debris have been exacerbated by several events in the last two years. Two events that have generated many thousands of debris elements have served to generate new concerns. The first event was the Chinese testing of a missile defense system to destroy a defunct Chinese meteorological satellite. This "test" created several thousand new pieces of orbital debris and was the single-largest source of space debris in recent years. The event was noted with particular concern

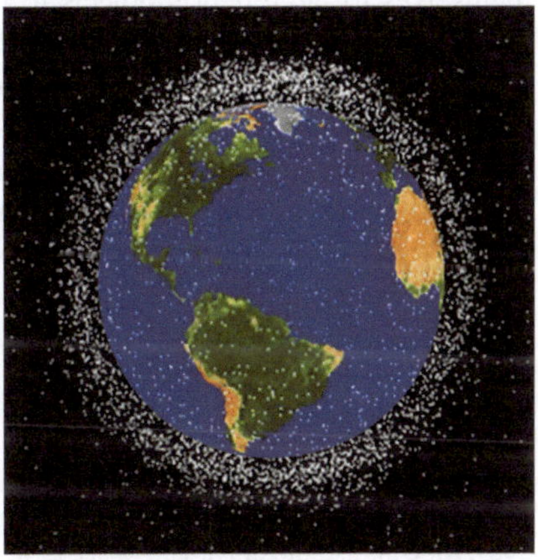

Fig. 3.5 Graphic representing orbital debris in low Earth orbit. (Graphic courtesy of the McGill Air and Space Law Institute.)

[8] *Op Cit*, Dennis S. Roddy, pp 177–181.

by the world space community because it was a deliberate and planned activity as opposed to a random act that is normally the cause of debris proliferation.

The other recent key event was the collision of an Iridium satellite and a Russian Kosmos satellite that also created several thousand new debris elements. There is now concern that unless significant new reforms are implemented there could be a further increase in debris that continues to multiply due to a "cascade effect" whereby ongoing collisions among debris elements already in orbit could just continue to increase. The Sun-synchronous polar orbit is of particular concern with regard to this destructive cascade effect.

The U. N. Committee on the Peaceful Uses of Outer Space (COPUOS) has adopted voluntary procedures for countries to follow to minimize the spread of orbital debris. These procedures were actually developed by a working group consisting of eleven space agencies known as the Inter-Agency Space Debris Coordinating Committee (IADC).[9] The COPUOS has also instituted a new "sustainability of space" study process that will explore possible ways to mitigate the amount of orbital debris. Also there is an association of satellite communications operators that has created an International Data Association (IDA) that provides real-time data as to possible "conjunctions" between operational, spare or defunct satellites. At this time only a limited number of operators that utilize GEO such as Inmarsat, Intelsat and SES, are participating in the IDA, but the number of participants continues to expand.[10]

The growing consensus view among experts in this area is that an active process to remove unwanted orbital debris will be needed to ensure the long term viability of key orbits now used for satellite communications. A wide variety of techniques to "remediate" (i.e., de-orbit) orbital debris have been suggested.

Possible remediation techniques that have been proposed for addressing space debris include the following: (i) ground-based lasers. These laser systems would fire pulses at large debris objects and alter their orbit to bring down derelict satellites and other large pieces; (ii) solar sail devices. These solar sail arrays would attach themselves to large debris objects to facilitate their de-orbit; (iii) tether-deployed nets. This system would deploy "nets" around space debris and speed up their de-orbit. (This system has been called "Rustler" for "Round Up of Space Trash – Low Earth orbit Remediation"); (iv) space mist. Satellites would be deployed in low Earth orbit that could spray gas mists and the frozen gas mist would serve to bring down orbital debris; (v) robotic systems. Robots would clamp on to space debris and then essentially throw the object into an orbit that would rapidly degrade; (vi) adhesives. In this approach very sticky adhesive balls composed of substances such as resins or aerogels would be "shot" at large space debris so as to alter their orbits and to bring them down over time.[11]

---

[9] The Inter-Agency Space Debris Coordinating Committee, http://www.iadc-online.org/inde/cgi

[10] Stephan Hobe and Jan H. Mey, International Interdisciplinary Congress on Space Debris, May 2009, McGill Institute of Air and Space Law, Montreal, Canada.

[11] International Interdisciplinary Congress on Space Debris, May 7–9, 2009 http://www.mcgill.ca/channels/events/item/?item_id=104375 also see David Kushner, "The Future of Space: Orbital Cleanup of Cluttered Space", Popular Science, August 2010, pp. 60–64

Unfortunately these approaches are all relatively expensive, at best only partially tested, and perhaps many years away from practical implementation. Also, most of these techniques essentially apply only to low Earth orbit. Fortunately this is indeed where the problem is most severe.

In addition to concerns as to the cost, long lead time to implement, and limited applications to only some orbits, there could be serious concerns with many of these techniques in terms of their being considered to be "space weapons." Systems such as high-powered land-based lasers that could serve to slow down and de-orbit space debris could also be applied for military purposes, such as to disable satellites.

The Design, Engineering, Manufacture, and Operation of a Communications Satellite

4

Introduction

The design, engineering, manufacture, deployment and operation of communications satellites constitute a complex and demanding process. The development of new and better technology has allowed satellites to improve over time. Today's satellites have much higher capacity, operate for a much more sustained period of time, operate over a broader range of frequencies, provide an ever-increasing range of diverse satellite services and work with more cost effective and user friendly ground antenna systems.

The Design and Engineering of the Communications Subsystem and Electronics

The essential element of a communications satellite is its communications subsystem. The communications payload includes the transmitting and receiving communications antennas, plus the associated electronics that filters out interference and intermodulation products, amplifies signals, modulates and demodulates multiplexed signals, codes and decodes digital information, and switches communications signals between uplink and downlink beams.

Satellite Communications Antennas

There are many types of antennas that can be used for satellite communications. These various forms of antennas range from rudimentary quite low gain to very large and sophisticated higher gain systems. In the earliest days of satellites the antenna could have been the simplest dipole antenna with a squinted beam, a very simplified conical antenna system as was deployed on the *Intelsat 1*, or "Early Bird" satellite. Communications satellites have since evolved from dipole and yagi arrays, to conical horn, spiral coil systems and three-axis stabilized parabolic antennas.

Today the satellites for mobile satellite communications are now deploying very large and high gain parabolic dishes that are 10 to even 20 m in diameter.

The gain of an antenna, as explained previously, is dependent on the square of the frequency (or the inverse square of the wavelength) as well as on the square of the antenna radii (i.e., the area represented by the antenna reflector). These two factors plus the "efficiency" of the antenna determine the "gain." The precise formula for calculating antenna gain is as follows:

$$G = \left(\frac{\pi d}{\lambda}\right)^2 \eta$$

In this formula G = Gain, d = the diameter of the antenna aperture, lamba is the wavelength associated with the radio frequency (RF) utilized and nu represents the efficiency of the antenna. (The antenna efficiency, typically around 65% to 70%, is determined by the accuracy of the shaping of the antenna reflector, the blockage of reception by the support structure of the feed system, etc.)[1]

The tremendous evolution in the sophistication of communications satellite antennas can thus be literally seen in the difference between the quite small and low gain squinted beam antenna of the Intelsat I in 1965 (See Fig. 4.1.) versus the gigantic antenna beam reflectors deployed by the Japanese ETS VIII satellite. (See Fig. 4.2.) This Japanese designed satellite was built to carry out experiments in mobile satellite communications. The objective was to test just how small and compact mobile satellite telephone transceivers for users on the ground could be if the satellite antennas up in space were sufficiently large. In other words the key to having ever smaller and more compact user handsets on the ground is most definitely driven by the need to create high density and concentrated beams that could more effectively irradiate power to Earth, especially from the great distance of GEO.

The tiny Early Bird antenna pictured above had quite a low gain that was many thousands of times less effective in performance than the very high gain and giant-sized Japanese ETS VIII deployable antenna system, pictured below, that was developed to support mobile communications satellite services. This is true even taking into account the higher frequencies bands utilized by the Early Bird satellite.

The key to success in developing very large deployable communications satellite antennas for mobile communications was not only in designing the large and low mass deployable satellite antenna, but also in developing the multi-beam feed system that can allow hundreds of different beams to be created by illuminating different spots on the antenna reflector. The multiple beam-forming feed system can thus create separate and isolated beams in various frequency bands so that the beams do not interfere with one another as they illuminate a different part of Earth or the ocean.

---

[1]Timothy Pratt and Charles W. Bostian, Satellite Communications, (1986) John Wiley and Sons, New York pp. 107–109.

Fig. 4.1 Small squinted conical beam antenna protrudes from top of Early Bird *(Intelsat 1)* satellite. (Graphic courtesy of the Boeing Corporation.)

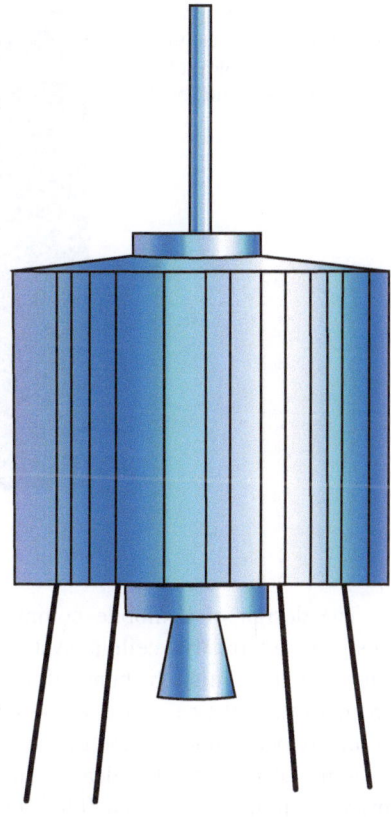

Fig. 4.2 Japanese ETS satellite with 19 X 17-m deployable antenna. (Photo courtesy of JAXA.)

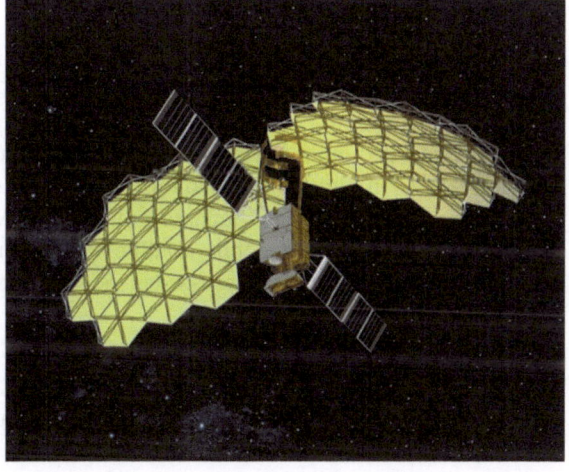

Fig. 4.3 *Inmarsat 4* provides
broadband mobile satellite
services via its deployable
antenna. (Photo courtesy of
Inmarsat.)

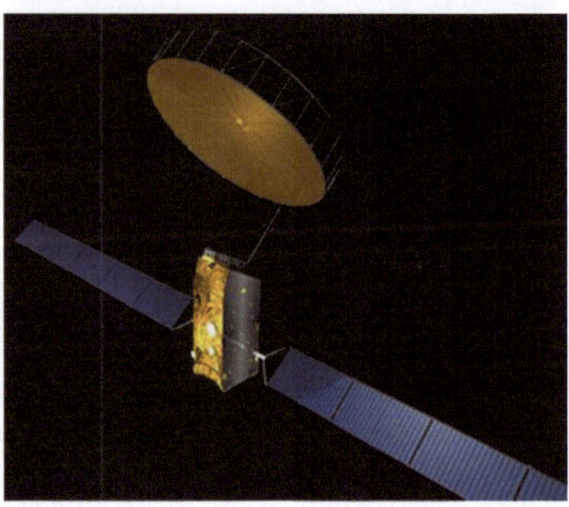

The design of a mobile communications system is conceptually much like designing a terrestrial cellular system. The very large difference is that in the satellite world there might be hundreds of beams covering a huge part of Earth's surface. In contrast, in the case of a terrestrial cellular system only a few cells are beamed from any one cell tower and the coverage areas are much, much smaller. Although in terms of concept the idea is almost identical, the technical challenge for the mobile communications satellite is much greater. This is because the beam-forming of hundreds of beams off of a huge antenna is much more difficult, the deployment of a very large reflector in space is a very hard task, and the satellite has to operate for a great length of time (i.e., 15 to 18 years) without any human engineer being able to make adjustments or provide maintenance.[2]

The currently operational *Inmarsat 4* satellite pictured below (See Fig. 4.3.) provides so-called broadband (BGAN) land, maritime and aeronautical mobile services for nearly one-third of the world's surface. This satellite is able to accomplish this vast coverage by creating and interconnecting over 220 very high-powered spot beams plus about twenty regional and "global" beams. These beams are formed via a very sophisticated multi-beam feed system illuminated 12 m deployable antenna system (pictured in Fig. 4.3) so as to form all of these separate beams. The number of beams formed by the Light Squared and Terrestar Mobile Satellite Systems (for hybrid terrestrial and land mobile satellite services in the United States) is even larger. These satellites will be discussed in the section on future market and technology trends.

---

[2] Denny S. Roddy, Satellite Communications, (2001) 3rd Edition, McGraw Hill, New York, pp. 128–133.

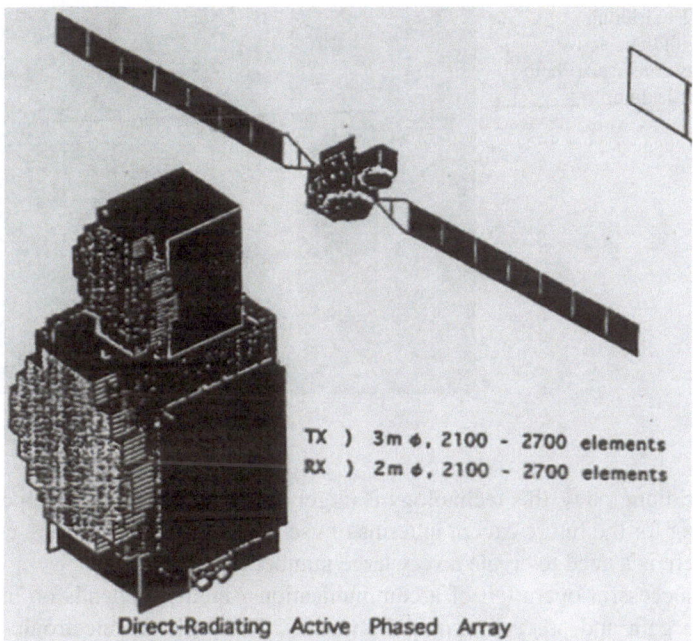

TX) 3m φ, 2100 - 2700 elements
RX) 2m φ, 2100 - 2700 elements

Direct-Radiating Active Phased Array

Fig. 4.4 Initial design for the Japanese gigabit experimental broadband satellite with a very large number of antenna elements. (Graphic courtesy of Jaxa.)

The latest innovation in satellite antenna technology is the "phased array" antenna. This technology depends on the putting an "array" of small electronic antennas together in a fixed pattern. This array of small antennas allows the forming of electronic beams. Instead of having a physically formed beam by illuminating a parabolic reflector, a phased array antenna can create a variety of "virtual beams" electronically. These phased array antennas are expensive to design and build. Thus satellite manufacturers today generally find that it is most economical to deploy large-scale passive parabolic reflectors as the basis for truly high-gain satellite antennas. There have been experimental active phased array antenna systems such as that deployed on the Japanese geosynchronous experimental broadband fixed communications satellite (see Fig. 4.4).[3]

The Iridium low Earth orbit constellation also used three phased array antenna panels to create a 48 beam pattern. Horn-shaped antennas also provided inter-satellite links to four satellites (i.e., links to the two satellites on each side and to the two satellite before and after in the near polar orbit) (see Fig. 4.5). Despite the operational

---

[3] Mark Chartrand, Satellite Communications for the Nonspecialist, SPIE, Bellingham, Washington, USA. P. 265. Also see Mark Williamson, *The Communications Satellite*, (1990) Adam-Hilger, London, U.K. pp. 178–179.

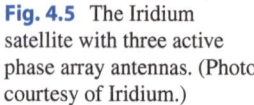
Fig. 4.5 The Iridium
satellite with three active
phase array antennas. (Photo
courtesy of Iridium.)

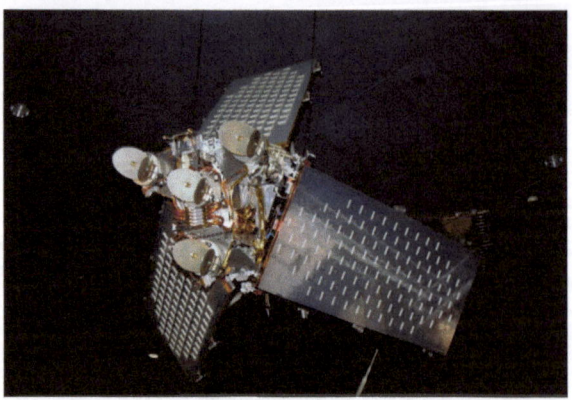

use by Iridium today, this technology is largely considered by satellite operators as
something for the future except in terms of use in beam-forming feeds – especially
when there is a need to create a very large number of beams.

The successful operation of a communications satellite depends on more than
just high gain and precisely pointed antennas. The on-board electronics is what
makes the antennas function properly. A lot of activity goes on onboard the satellite,
and over time more and more functionality has been added so that today, the onboard
electronics package carries out the following functions:
- receiving of the uplink signal through the feed system once the signal has been
  concentrated via the uplink antenna reflector;
- filtering of the uplink signal to largely eliminate unwanted interference and
  inter-modulation products. Special polarization "filters" also separate the "vertical signal from the 90-degrees-out-of -phase "horizontal" signal in the case of
  orthogonal polarization. In the case of circular polarization the polarization filters
  separate the left hand circular waveform from the right hand circular waveform;
- amplification of the uplink signal to make it stronger;
- translation of the uplink signal to a downlink frequency;
- re-amplification of the downlink signal;
- switching of the downlink signal to the proper downlink beam via an onboard
  switch; and
- transmission of the downlink signal through the feed system to be irradiated off
  of the downlink antenna reflector.

The next section briefly describes the design and engineering of the various electronic components of the satellite communications payload.

The Filters and Dual Polarization Systems

Electronic filters serve to divide the "wanted signal" from the "unwanted signal" in
adjacent frequency bands of spurious signals to or from adjacent communications

satellites. These filters, particularly in the case of digital signals, perform quite well in stripping off interfering signals or electronic noise.

As noted earlier, In many satellite systems there is also the issue of polarized signals to be considered. In order to double the available spectrum that can be used to relay signals via satellite, dual polarization of the signal waveform is often employed.

The function of the polarizers at the satellite and at the Earth station is to sort out or "filter out" the signals from one another so that the same spectrum band can be utilized twice to send intelligible signals. Since the available spectrum bands are typically only 500MHz or 1,000 MHz across, dual polarization serves to effectively expand the 500 MHz band, so it seems to be the equivalent of 1,000 MHz and to expand the 1,000 MHz band so that it seems to be the equivalent of 2,000 MHz.

In the case of orthogonal dual polarization the signals are 90 degrees out of phase and therefore distinguishable from one another. In the case of circular polarization the waveforms are either moving in a circular pathway that is rotating either in a left hand circular motion or in the opposite right hand circular direction. The "wanted" polarized signal is typical 20 to 30 dB stronger than the "unwanted signal" in each of the polarized receivers. This means that in each of the two polarized receivers the polarized signals that are being discriminated from each other seem to be 100 to 1,000 times greater in intensity than the oppositely polarized signal.

Amplifiers

Once the filters and the polarizers have weeded out the interfering signals and separated the dual polarized signals from one another, the next step is to boost the intensity of the signal by the use of an amplifier. In the original communications satellites this amplification was accomplished by a device called a traveling wave tube amplifier that is often abbreviated to TWTA. Today in the major bands used for satellite communications, namely C band (6 and 4 GHz) and Ku band (14 and 12 GHz) TWTAs have been largely replaced by solid state power amplifiers (often referred to as SSPAs) that are much less massive and have a longer effective lifetime. In the extremely high frequency bands, and especially in the case of the Ka band or for direct broadcast satellites, TWTAs are still used, especially above 30 watt power levels, because it is still largely not possible to generate the high power needed to support communications in these bands with solid state technology and thus tube technology is still employed.

Digital Modulation

How does information of all types such as voice conversations, numbers, video images, and music get converted from their original form into radio wave signals that can be relayed via satellite to a distant location and then converted back to the original information? To many people this process actually seems as if it is magic.

Actually it all depends on the physics of electronics plus a mathematical encoding process that lets electrical voltage information represent a code for different types of information. The original information, one way or another, has to be "encoded" to convert the "source information" into some sort of code that represents the initial sound or image.

The next step involves some sort of "modulation" or varying of a signal to change the characteristic of the radio wave. The main options are to change the radio wave's intensity (i.e., amplitude height), its frequency, or its phase (i.e., whether the radio wave vertical, horizontal, at a 45 degree angle, etc.) over time in order to be able to transmit that code information over a radio carrier wave. This will be explained in a bit more detail and hopefully with some clarity a bit later.

The simplest form that was used to take "source information" and encode it into a written message that could be sent as an electronic signal was employed in the 19th century via the Morse code. In this case the letters of a word plus the "stop" between words was converted into a series of short dots and longer dashes using a code developed by Samuel F. B. Morse. This process worked to send messages over a telegraph wire. The idea that messages could be sent in this manner was initially denounced in the U. S. Congress as "black magic." This process of encoding and sending information electronically via Morse code, however, was slow in comparison to modern techniques, and in the process mistakes could easily be made.

Over time clever inventors have found more and more efficient ways to encode information. Today this typically involves converting information into a digital code. The key is to rapidly sample changing source information such as speech or a television transmission and convert it into a digitally coded electronic form such as a voltage level that can be decoded at the other end to exactly represent the same information. Digital encoding can be quite efficient in that 256 different levels of sampled information can represent different things in binary code by using just eight transmitted bits. A transmitted bit can represent either a zero or a one. If we recall that 2 to the eighth power is 256 then it becomes clear that the various combinations of transmitted zeroes and ones contained within just 8 bits can represent a gradation of 256 different things in terms of frequency, volume, etc.

The information that we start with such as speech is a continuous stream of information. This "analog" information is converted into a series of discrete and discontinuous information as digitally coded "samples." The question that naturally occurs to anyone thinking about this is how many samples must be taken to get a good quality representation of the "source information." A very clever engineer at Bell Labs named Harry Nyquist determined that if the sampling and encoding of information occurred at a rate that was at least twice the rate that information is itself changing then the information can be captured with true fidelity.

The source information is sampled and digitally encoded through the "modulation" or changing of a radio wave signal to create a digital signal that can be efficiently sent and received thousands of kilometers away. The initial sampling and modulation of an electronic signal is accomplished within the baseband frequencies associated with that of normal speech.

There are three types of modulation commonly used in telecommunications. These are as follows: (i) amplitude modulation, where the power is increased or decreased to create a so-called analog signal that mirrors the original sound or image; (ii) frequency modulation, where the power or amplitude remains constant and the frequency is varied to create the analog signal, and (iii) phase or angle modulation. Phase modulation and particularly "phase shift keying" (PSK) is most commonly used in digital modulation to create a digitally encoded signal via sampling techniques. PSK allows the encoding to occur to represent a string of binary numbers that is a "digital model" of the original signal.

The technique known as PSK sounds highly technical, but the process is actually quite simple. If there is a two phase shift the code is reflected in terms of the signal wave being at either 0° or at the opposite angle of 180°. To create the "digital code" we simply let 0° represent a "zero" and 180° represent a one.

One can start to be even more clever and say pack twice as much digital information into the encoded information by refining the various possible shifts of a wave pattern to convey more binary numbers via phase shift keying. Instead of bi-phase shift keying (BPSK) where we shift only between 0° and 180° we could use four phases, or so-called quadra-phase shift keying (QPSK). In this scheme "00" would be at a 0° phase shift. Then "01" would be at a 90° of phase shift, while"10" would be at 180° of phase shift, and "11" would be at a 270° of phase shift. One might then jump ahead and not stop there. We could clearly do this again and pack in much more information again if we went to an 8-phase shift keyed system. In this approach we could let 0°, 45°, 90°, 135°, 180°, 225°, 270 ° and 315° of phase shift represent "000", "001", "010", "011", "100", "101", "110" and "111". The problem is that there are limits to just how far one can go in "packing information" into a signal using the phase shift keying approach. This problem particularly applies to the field of satellite communications.

In the field of fiber optic transmission the quality of signal is very high and the potential interference very low – almost none in fact. In the world of digital communications interference or noise is measured in terms of a so called "bit error rate." In the field of fiber optic transmission the bit error rate might be something like 10-12. This is like saying that there would be only one bit error in every trillion bits of information sent. In the world of satellite communications in clear sky conditions very low bit error rates – or very low interference levels – can be achieved. In the case of heavy rain conditions performance deteriorates.

In the case of what is called high levels of rain attenuation, interference or noise increases, and it becomes more difficult to differentiate 0° from 45° with a high degree of accuracy. This is to say that in the world of fiber optics one might even use 16- Phase Shift Keyed encoding, but one would probably not do this in the world of satellite communications. It might be the case that one could use something like 8-phase encoding of information (i.e., 8-PSK) but also have the capability to drop back to Bi-Phase Shift Keyed (BPSK) or Quadra-Phase Shift Keyed (QPSK) encoding in satellite beams where heavy rain or adverse atmospheric conditions started to degrade performance. There are techniques to correct for bit errors, and this technique is called "forward error correction" (FEC), but this serves to slow down transmission speeds and thus it is best to have the highest transmission speeds possible consistent with high quality performance – or a low bit error rate.

There are a host of modulation and encoding techniques that can be used to improve the density of information that can be sent through a satellite via the available radio frequency spectrum that is available. The "typical" density or efficiency factor for today's communications satellites is ordinarily in the range of one bit per hertz of spectrum to about 2.4 bits per hertz. This efficiency is very important for satellites since there are only a few blocks of radio frequency spectra allocated to satellite communications, most notably in the C band (6 and 4 GHz), Ku band (14 and 12 GHz) and Ka band (30 and 20 GHz). These various advanced encoding schemes and compression schemes are discussed in the chapter on trends for the future.[4]

Once a signal is encoded at baseband it needs to be converted to the higher radio wave frequencies used in satellite communications just referred to above. Because the radio wave frequencies are so very high they are often converted from baseband to higher intermediate frequencies (IF) and then reconverted again to the radio waves (RF) that are used to uplink to the satellite. Once it arrives at the satellite, the signal is again converted to the downlink frequency and sent to the Earth station. The RF signal, once it reaches the end user, is again changed to baseband so that it is in the audible range that humans use for speech.

Multiplexing

Beyond the process of digital modulation in modern satellite systems there is also the process that is called multiplexing. Once the source modulation has occurred the next step is to combine that signal with others so as to efficiently transmit information through broader band satellite networks. This means multiplexing the signal into broadband carrier signals and then de-multiplexing the signal at the other end of the transmission. A variety of multiplexing schemes are used to combine signals. In a way this is like a high voltage electric transmission from which lower voltage household power is derived, but in reverse. A variety of techniques have been used for this purpose. The most common multiplexing/de-multiplexing (mux/demux) systems include the following as described in Table 4.1.

One of the most important milestones in the evolution of satellite communications was when satellite designers moved to the new practice of switching a signal from an uplink spot beam to a downlink spot beam. This shift occurred through the adoption of onboard switching that allowed the shift of a signal from one beam to another. Instead of a satellite transmission going up and down within a single global or zonal beam there was now complete flexibility that allowed a voice, radio, data

---

[4] For further information on modulation and encoding see: Mark R. Chartrand, *Satellite Communications for the Nonspecialist*, (2004), SPIE Press, Bellingham, Washington, USA pp. 103–118. Also Gary D. Gordon and Walter L. Morgan Principals of Communications Satellites, (1993) John Wiley and Sons, New York. p. 114–118

Table 4.1 Different types of multiplexing systems used in communications satellites beam switching

Various Types of Multiplexing Techniques Utilized in Satellite Communications	
Multiplexing Technique	Brief Description of the Multiplexing Technique
FDMA (Frequency Division Multiple Access)	This analog-based system that operates in the frequency domain combines baseband modulated signals via a multiplexer so that a broader band carrier waves can efficiently transmit the combined signals. This technique was widely used in the early days of satellite communications, but digital based multiplexing systems are now used instead.
TDMA (Time Division Multiple Access)	This is a very widely used digital multiplexing system that combines baseband signals into a combined carrier wave using the time domain.
CDMA (Code Division Multiple Access)	This is a very widely utilized multiplexing system that is particularly used with mobile satellite communications systems and cellular. The baseband signal is de-multiplexed by using both a digital time slot and a particular code for each signal.
Aloha	This is a system used for narrower band communications that was originally developed at the University of Hawaii. This system uses random transmissions in the time domain. In some cases there are simultaneous bursts that then require retransmission.
Slotted Aloha	This is a modified version of the Aloha system that is more structured to avoid conflicting transmission in the same time slot.

or television channel to be uplinked in one spot beam and then switched to another spot beam on board the satellite.

The greatly expanded use of spot beams and the ability to switch from one beam to another easily produced three key advantages. The first advantage was that the beam in the uplink or downlink could be much more tightly formed, and thus there was much less power loss due to beam spreading or so-called path loss. The second advantage was that the various beams could be separated in such a way to allow frequency re-use with much greater efficiency. This means that the more beams that were added – and the more tightly they were formed – the higher the power within the beam coverage area can be illuminated on Earth below. In short, more narrow spot beams of higher intensity allow for more frequency reuse.

Finally with the use of a carefully formed large multiple-beam parabolic reflector, these multiple beams could be formed using a multi-beam feed system that simply illuminated different parts of the same antenna reflector to create each beam. Thus one needs only one high gain antenna for the uplink reception and one high gain antenna for the downlink transmission. It is thus left to a complex feed system to transmit or receive a large number of beams via the large transmitting or receiving antenna reflectors.

There are, of course, disadvantages to operating with all these uplink and downlink beams. The design and creation of the multi-feed system, of course, becomes increasingly difficult and technically challenging. As the number of beams increases the switching system needs to interconnect a larger and larger number of beams. The number of possible interconnections increases exponentially. This also means, however, that the switch is more difficult to design and build and this drives up costs. Also the number of possible failures in a single point of failure device also increases rapidly. Further there has to be a more complex set of software to test and pinpoint a failure in a particular pathway. This means that there has to be complex fault detection software as well as artificially intelligent software to allow for rerouting of traffic on a suitable alternative pathway. The history of satellite communications can, in a sense, be plotted in the number of spot beams generated on a communications satellite. The number of pathways to be connected on a communications satellite has gone from two to thousands to virtually millions on the latest mobile communications satellite systems.

Onboard Processing

The addition to any communications satellite of an electronic switch to route traffic among a large number of spot beams, plus special software to assist with the instantaneous routing and fault detection, represented a major step into greater complexity. Sophisticated software involving a considerable amount of code to allow very rapid fault detection was suddenly necessary. If a particular switch could malfunction at any time, then this required adding considerable additional intelligence to the satellite in terms of a high speed processor, a backup processor, and a complicated set of diagnostic instructions to see where the fault might lie. In a six by six switch for instance a total of 360 pathways are possible.

Up until this time all of the real intelligence and computer capability largely existed on the ground. Satellite engineers began to think in terms of satellite architecture, where more "intelligence," switching capacity and signal processing might take place on the satellite in space rather than on the ground. There are a number of reasons why one might make the satellites in space "smarter" and more capable. The number one reason is that if a satellite signal were broken back down to the baseband and the equivalent of the original signal regenerated, this could help a great deal with overcoming problems of interference and rain attenuation. Rain attenuation is most severe in the uplink because the uplinked signal is at the higher frequency. If one could essentially recreate the original signal and virtually wipe out the effects of interference that occurred in the uplink signal than slightly more than half of the problem could be eliminated. Hybrid modulation and multiplexing schemes and more sophisticated encoding schemes could also be employed if there is on-board processing (OBP) capability onboard.

Today a number of defense-related satellites have employed onboard processing, in some instances because this additional capability was considered to be strategically important. The Iridium satellite system used onboard processing, and

Table 4.2 The pros and cons of onboard processing. (Table prepared by the author.)

Strengths and Weaknesses of Onboard Processing (OBP) for Communications Satellites	
Pros	Cons
OBP at baseband level could greatly serve to overcome rain attenuation.	Addition of OBP could add greatly to cost and increase satellite mass.
OBP can contribute to more efficient encoding, modulation and multi-plexing schemes by using different systems for the uplink and downlink.	Addition of OBP adds to the complexity of the satellite that can delay manufacturing and advance the projected mean time to failure (MTTF)
OBP can add to quality of service and reduce bit error rates (BER).	This technology represents state-of-the-art systems. This means a system that has not been as thoroughly tested and its reliability fully proven. In short use of this technology represents many unknowns.
OBP can serve to accelerate Internet related throughput and assist with problems associated with transmission delay	Investment in OBP adds more mass and volume to the spacecraft. This represents an opportunity cost in terms of not being able to add more fuel or larger batteries.
Upgraded performance via the use of OBP could allow the use of smaller aperture user antennas in key instances where use of such micro-terminals might be essential.	OBP presents more critical points of possible total system failure

this system now has demonstrated a collective lifetime of over 600 satellite years of in-orbit operations. As the use of OBP capabilities become more common and the technology more mature, the elements associated with cost, longer manufacturing time, and/or additional weight or volume should shrink in significance and the use of the technology will increase. This is by no means the guaranteed longer-term trend. In cases of hybrid systems for mobile communications such as Terrestar and Light Squared, the processing has largely returned to ground-based systems. Table 4.2 provides an assessment of the "pros" and cons" of on-board processing.

TTC&M

The design, manufacture and test of the tracking, telemetry, command and monitoring (TTC&M) system for a communications satellite is extremely important. If the TTC&M does not work then the satellite network cannot operate and the mission is a total failure. The TTC&M subsystem, in most cases, uses different frequencies and operates through a separate antenna system for purposes of redundancy. The ground systems also typically use separate ground antennas for these functions as well, and the TTC&M functions are usually coordinated and controlled through a centralized Satellite Control Center (SCC). This independent line of communications and functionality is separate from the actual commercial communications links provided by satellite networks for a variety reasons. The objective is to increase reliability, provide greater opportunity to recover the satellite if it should lose orientation pointing away from Earth, and to separate the communications function from the TTC&M activity so that vital control of a satellite during a major emergency or even a temporary outage can be given top and focused attention. Part of this strategy is to have as much redundancy and "fail safe" capabilities built into the TTC&M function as possible. This means that at least two or more stations are able to provide TTC&M services to a satellite at all times. It can also mean that separate facilities can be used to authenticate commands so that a satellite system can be protected against hackers or anyone seeking to disrupt or disable satellite operations.

Tracking: The tracking of a satellite is simply to be able to know where it is. The exact tracking function becomes more difficult as the satellite moves in higher and higher orbits and thus further away from Earth and moves at varying speeds – especially if it is in an elliptical orbit. A GEO satellite is, of course, a special case. It is obviously not difficult to track a satellite in geosynchronous orbit in terms of determining its "approximate location," and its angular velocity is constant. This is because the satellite appears to be stationary with regard to the ground and thus rapid tracking by a TTC&M ground antenna is not required. Nevertheless it is hard to locate exactly where a geosynchronous satellite is when it is in fact almost a tenth of the way to the Moon. In general, one requires a highly focused larger aperture tracking station that uses higher frequencies to track a satellite in a very high orbit and vice versa. One can use radar or radio signals from the satellite to track a satellite's orbit in terms of angular velocity and altitude. The satellite itself can use a process called triangulation to determine its exact location at any one point in time. This involves a calculation based on onboard sensors "perceiving" where the satellite is in relationship to Earth, the Sun or a particular star. A more common way today is to track a beacon signal sent up from a TTC&M Earth station at a precisely known position. One can also use the Global Positioning Satellite (GPS) system to determine a satellite's location. However, this system is not as accurate in space as it is on Earth's surface.

Telemetry: This is a means for the Spacecraft Control Center (SCC) to obtain essentially real-time information about the "health" of the satellite. A satellite is equipped with a number of telemetry devices that measure current conditions in the satellite and then relay that information to ground controllers. These monitoring devices measure electrical currents within the satellite, the power generated by solar arrays, the depth of discharge on the satellite batteries, the temperature within the spacecraft bus, or whether a satellite amplifier, filter, repeating device, switch, or processor has failed. The great bulk of telemetry data flows to the SCC as a continuous stream. This information is registered and recorded by computers. And as long as this data stays within fixed limits or does not indicate a rapid change in conditions everyone is relaxed and knows the satellite is doing its job well and reliably. But when an alarm rings indicating that something is out of whack, the ground controllers come running to see what the problem might be. If there is a problem indicating that the satellite is not tracking the correct orbit or one of the many set levels for voltages, temperature, power generation, etc., are out of normal operating specs, then the SCC comes alive. There may be rapid response "fixes" such as switching from a repeater that has failed to a backup spare component.

There may be a need, however, for diagnostic tests. Spacecraft engineers on call will likely be consulted to assess the nature of the problem and the most effective solution. The onboard switches are critical to shifting from a failed component to a backup or to route around a failed circuit. The restoration procedures have evolved to a high degree of sophistication after nearly a half century of in-orbit experience. These procedures have been reduced to algorithms, and this has allowed some spacecraft to feed their telemetry data into an onboard computer to allow the satellite to engage in "autonomous operation." In most instances there is an override capability to release remedial response from the onboard computer to on-the-ground controllers.

Communication satellites have increased in complexity, size and cost to manufacture and launch. This means that the number of onboard sensors, monitors, pressure gauges, and network switch diagnostics have increased over time. This consequently means that the amount of telemetry readout has increased considerably for the large, complex satellites, and the bandwidth associated with telemetry has also increased to send the data to the SCC in near real time.

The Satellite Design and Engineering Process

The design and engineering of a communications satellite starts with the specifications of the desired performance of the spacecraft. This begins with basic parameters such the radio frequency spectra to be utilized, the desired digital throughput capability, the performance characteristics of the user antennas, and the orbital location of the GEO satellite and the desired service area or if appropriate the intended LEO or MEO constellation as well as the desired lifetime of the spacecraft. In many cases the satellite operation will do system optimization studies to see which of these key variables might be altered so as to optimize performance and to minimize costs.

When the Iridium satellite network was first designed as an LEO constellation the initial design called for a symmetrical polar orbiting system that allowed four inter-satellite cross links on each satellite. This allowed connectivity between satellites before and after as well as to each side. Such cross link capabilities allowed for the TTC&M network to be much simpler. Since telemetry, tracking and command information could be relayed between and among the satellites far fewer ground stations could be built to perform this function.

After further systems analysis the original constellation of eleven satellites deployed in each of seven different polar orbits (i.e., a constellation of 77 satellites plus spares) was revised. The revised plan was to build more powerful satellites with more spot beams to increase link margins. In the revised design each satellite had 48 spot beams rather than 37. Also the constellation was revised so as to deploy eleven satellites in six polar orbits (i.e., a 66-satellite constellation plus spares). It was calculated that the satellite constellation could be built and launched into orbit for a similar total cost but that the performance would be significantly better in terms of equivalent isotropic radiated power (EIRP).

When the Globalstar mobile satellite constellation was designed, close to the same time a number of different design parameters were considered. The designers of the Globalstar network decided to build a constellation with fewer satellites in orbit (i.e., 48 satellites plus spares). The designers considered that there would be very little service demand in the polar region and decided to deploy a constellation that was not able to provide service above 70 degrees north or south latitude and also to launch the constellation at a higher orbital elevation so that each satellite could have a wider range of coverage. The thought was that this design would cover virtually all of the needed service area and would require the design, manufacture, testing and launch of fewer satellites. It was recognized that there would need to be far more TTC&M Earth stations deployed to operate the constellation successfully, but the elimination of the cross links freed up additional space, mass and power to use for the prime communications mission. Further it was thought that various partners around the world would build and cover the cost of the TTC&M stations while the consortium was paying the costs of building and launching the satellites.

Had the projected global market for satellite mobile services materialized as projected by respected international consulting firms, both of these multi-billion dollar (in U. S. dollars) satellite systems would have likely had proved economical. When the total costs for the design, manufacturing, testing, launch, and operation of the two systems were ultimately computed (i.e., the total cost of the satellites, the launches, the launch insurance, the gateway and TTC&M stations, the user terminals and the system operation) the difference was really not very great in terms of a percentage. The design and system engineering choices for Iridium and Globalstar were dramatically different. Yet, the financial results turned out to be only marginally different in terms of total capital costs (i.e., both systems added up to be about $5 billion (in U. S. dollars) when total space and ground systems costs are considered). The Iridium system offered somewhat higher capability and greater Earth surface coverage at somewhat higher cost. Iridium involved higher spacecraft costs but lower TTC&M costs on the ground.[5]

The real point to conclude from this analysis is that there are almost an infinite number of tradeoff choices in a system optimization study for a satellite system that includes the following factors: orbital configuration or constellation, service coverage, lifetime, link budget and margin, user transceivers and antennas, spacecraft antenna design, testing processes, etc., and there is no necessarily right or wrong answer as to design. Ultimately the key to success and/or failure is in accurately projecting market demand and service revenues versus the cost of deploying and operating a satellite system and associated user equipment. In the case of Iridium and Globalstar the projections as to market demand proved to be off by at least two orders of magnitude, and both systems ultimately ended as market failures.

The system design and optimization process for GEO satellite systems tend to be much simpler than is the case for LEO or MEO constellations. This is in part because most GEO systems, particularly for fixed satellite services, have evolved over time, and there is a well- established history of market demand and market growth. Many

---

[5] Joseph N. Pelton, Satellite Communications 2001: The Transition to the Mass Consumer Markets, Technologies and Systems, (2001) International Engineering Consortium, Chicago, Illinois, pp. 88–93.

domestic satellite systems have evolved over the years and many of these started on the basis of leased capacity from large global satellite systems. Further, GEO systems can be established on the basis of one operational satellite and one launch. This makes the entire design, engineering, manufacture, testing and launch much easier from start to finish. This is particularly true if the satellite operator can choose a manufacturer who has previously built and deployed a spacecraft platform of very similar design, and the same is true for the satellite antenna and communications payload.

Today there are less than a dozen spacecraft manufacturers and satellite system integrators that supply most of the world's communications satellites and the same is largely true for Earth station antenna systems and mobile communications satellite transceivers and handheld units. These organizations have developed a reliable and well-known range of products that have been proven over time. Satellite operators, when they develop performance specifications for spacecraft, for TTC&M or gateway stations or for user antennas or handsets, for the most part, have a good starting point from which to specify the next generation satellites.

There are certain new areas that represent the next frontiers that will be discussed in the section on future satellite systems. These are areas such as the very highest frequencies, the latest capabilities in terms of onboard processing and signal regeneration, or the latest in beam-forming and phased-array antenna systems. Most satellite operators are not pushing the envelope of the very latest technology and can safely specify spacecraft with a high degree of confidence that their needs can be met at predictable costs.

Satellite Manufacture and Testing

The first satellites were simple devices that were small in mass and volume and involved fairly uncomplicated electronics and antenna design. These could be manufactured fairly quickly, tested and launched in a fairly short period of time. The greatest challenge, in these early days was a reliable launch. Over time the satellite designs have become more complicated as the power increased from a few hundred watts to as high as 12,000 to 15,000 watts, and satellite antennas increased from a very small omni or conic antenna to huge parabolic reflectors that can have an aperture diameter of up to 20 m. These complex spacecraft with thousands of parts and perhaps a combination of solid state and traveling wave tube technology became increasingly hard to design, manufacture and test for reliability. Inertia wheels (and today most commonly high speed reaction wheels), advanced battery designs, heat pipes, and new carbon-epoxy construction materials also added to the complexity of design, manufacture and testing. The time to design, build and test a satellite that began in time periods measured less than a year stretched to three years, and for the most complicated satellites to four years or even longer.[6]

[6]Gary D. Gordon and Walter Morgan, *The Principles of Satellite Communications* (1993) John Wiley and Sons, New York. pp. 442–464.

The long lead time for setting performance standards for a satellite, receiving bids from manufacturers, choosing the winning design, manufacturing the satellite and testing it so that it was ready for launch has increasingly served to put satellite operators at a strategic disadvantage *vis a vis* terrestrial telecommunications technology. This has led to attempts to speed up the manufacturing and testing of satellites by suppliers. This has been accomplished in a variety of ways that has been different for satellites designed for GEO versus satellites planned to be deployed as part of an LEO or MEO constellation.

In the case of GEO satellites, where far fewer satellites are involved, the stress has been on having key component systems such as satellite antennas, spacecraft platforms and power systems available for assembly and testing on an accelerated basis. This process reduces the ability to obtain highly customized spacecraft tailored to a satellite operator's particular systems needs, but it can obviously speed up delivery and reduce engineering, manufacturing, testing and labor costs. Components and subsystems that have previously been manufactured, proven in space and prequalified through preliminary testing add reliability and reduce costs.

The approach with regard to multiple satellites for LEO or MEO constellations involve the concept of trying to develop precise high quality manufacturing techniques and component level testing on an assembly line basis that increases in speed and quality as the program reaches increasing levels of maturity. In the case of the Iridium constellation some 100 satellites were built and tested, and in the case of the Globalstar satellite systems, some 60 spacecraft were manufactured and tested. Throughout this process, the production and subsystem testing increased in speed over time. At the end of the Iridium production schedule satellites were being built and rendered qualified for launch in only a four-and-half-day period.

The idea, in the case of large production runs for satellite constellations, was to employ high quality production techniques and prequalification of components to the highest possible quality (of standards so that the spacecraft could be produced almost as if they were VCRs or television sets). There was concern that such a production and testing process put enormous pressure on the subsystem testing and limited integrated testing of the complete spacecraft. There were some initial systems failures with some of the earliest flight models, including operational "cockpit errors" (i.e., errors by technicians providing network control for the spacecraft constellation in real time). These problems were largely overcome with component redesign and better system control procedures and better training for those managing the network. In practice, from 1997 through 2010 the Iridium satellite system has proven extremely reliable, with the fully deployed system achieving a combined reliability record of over 600 satellite years in orbit – a new world record for the longevity of an LEO constellation. The Globalstar system likewise achieved a remarkably high degree of reliable performance. The longevity and reliability of these satellites were attained, although both systems were subjected to accelerated testing and validation processes. (See Fig. 4.6.)[7]

---

[7]Ibid, p. 162 and p. 443.

Fig. 4.6 The Globalstar satellite manufactured with accelerated production and testing processes. (Graphic courtesy of Globalstar.)

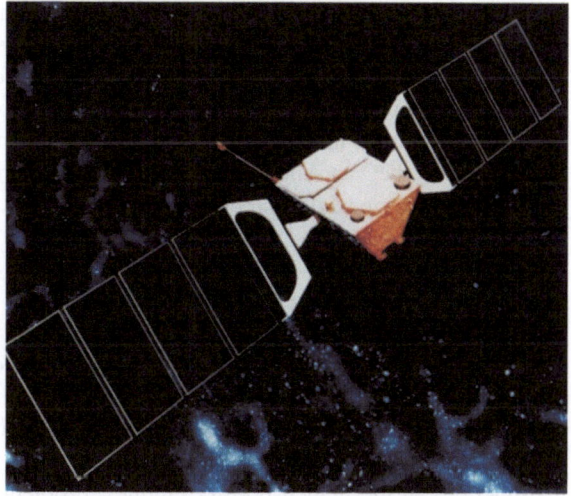

The testing of spacecraft is obviously important, but it should be recalled that there are two aspects of the testing process. One is to assure the satellite engineers that the "performance" of the spacecraft in terms of power (i.e., EIRP), beam coverage and contour, and filters and amplifiers meet or exceed all specifications. The other testing objective is to seek the maximum reliability and lifetime. The performance testing is most often accomplished by so-called "near and far range testing" of the satellite's antennas, where tests are undertaken with regard to gain, beam shaping, etc. This is only a simulation of actual performance in space, but antennas within the test range can give a fairly accurate indication of transmission capabilities. There are also tests within special chambers that have no radio echoes (called anechoic chambers) to assess reception and electronic performance.

The reliability testing varies from satellite program to satellite program. One of the more common tests involve shaker tables that simulate the stress and strains of satellite launch and special concerns such as the so-called pogo effect, and low level vibrations and their harmonics. Another common test involves thermal vacuum chamber testing. This simulates the conditions that spacecraft will encounter in outer space. Such tests may last for days at a time, but this of course cannot duplicate the conditions that spacecraft will endure for perhaps 15 to 18 years of operation in space and going through scores of eclipse seasons over time.

Many of the tests involve particular components and subsystems. Batteries and solar cells are tested for durability. In some cases heat or other stress elements are added to simulate extended periods of time. Transponders, antennas, heat pipes, thrusters, momentum wheels, power converters, even application-specific integrated circuits are subjected to lifetime and stress tests. Satellite design engineers seek to use component parts that have a demonstrated reliability and many years of in-orbit experience. Unfortunately it may take many years for a problem due to faulty manufacturing process or design weakness to show up as a reliability issue. There are examples such as planar integrated circuits that have manufacturing problems with impurities that were not detected that led to growth of spurs that resulted in component failures years after these units were launched into orbit. There are also attempts to "design in" protective systems. Solar arrays that are designed for satellites that are to be deployed in medium Earth orbit and thus going to have higher exposure to radiation from the Van Allen belts will be coated with an additional glass layer of shielding to provide additional protection for the solar cells.

Satellite Deployment and Operation

The launch and deployment check-out of communications satellites and their operation represents a critical step in the success of a satellite network. The launch of a communications satellite is the first key step. Launch operations are today more reliable than in the past. Nevertheless there can still be launch failures. These can be malfunctions in the launch booster engines, problems with stage separation, and even software errors. One launch failure came from the fact that the software on a launcher was set for the deployment of two smaller satellites rather than one large spacecraft. The most common launch failure today typically comes from under performance of the rocket booster and thus the achievement of the wrong orbit or the rocket motor exploding. Special care must also be taken to protect against severe vibration or "pogo-effects" that disables or creates malfunction in the satellite. Another critical step comes when the satellite's antenna and solar arrays are deployed. As these parts of a satellite have grown larger and larger, the danger of improper deployment has also increased. The Light Squared satellite with a 22 m antenna launched in late 2010 for instance had major deployment problems. Only "jostling" the spacecraft by rhythmic firing of thrusters allowed this huge antenna to finally fully deploy.

Satellite Sparing and Restoration of Satellite and Terrestrial Service

Despite the extensive reliability testing that is undertaken to improve component and systems design, these actually are no guarantee that there will not be failures in the hostile environment of outer space. Clearly repair of in-orbit satellites by astronaut service personnel is neither technically or economically viable. In short, reliability testing can only provide a higher level of confidence against satellite failure. A certain percentage of satellite failures can be expected. About 10% of commercial launches of unmanned satellites can be expected to be failures. A certain number of satellite components can also be expected to fail, and if that component or subsystem represents a so-called single point of failure then the satellite also becomes a total loss.

The prime strategy to combat in-orbit failures is thus "sparing" of key components and subsystems. Beyond the sparing of components one can also even provide for a complete satellite spare. This can be a "live" in-orbit spare, ready for immediate switch over to operation that is co-located with a geosynchronous satellite in essentially the same orbit slot. This "spare in orbit" can be repositioned to an appropriate orbital position to restore service within a matter of days or weeks. Or it can be an "on the ground" spare that is launched as soon as possible. The U. S. military is currently developing an automated space plane that can be mounted to an Atlas V or a Falcon X launch vehicle to allow a very rapid deployment of a tactical communications satellite at a strategic location.

Satellite manufacturers are also developing new techniques to be able to assemble and launch a replacement or urgently needed satellite on an accelerated basis. In this case the satellite manufacturer assembles key elements of an entire spacecraft (i.e., antenna reflectors, transponders, batteries, solar array, and spacecraft bus components). This allows the manufacturer on demand to assemble and test a "generic satellite" within a few months time for rapid response time launch. Such a "generic satellite" may not have optimum capabilities for a particular mission, but by the right assemblage of subsystems it may be able to be close to the desired specifications. This type of approach to rapid satellite manufacture is currently driven by military-related requirements, but if this approach is successful (i.e., in terms of reduced cost, reliable performance and accelerated deployment schedules) then commercial providers of satellite services or conventional governmental missions may also tend in this direction.

One of the primary strategies in designing satellites for reliability was to have backup satellite transponders. After some years of experience most satellite designers opted to have only one redundant transponder for every two transponders with the ability to switch flexibly to the backup component as needed. Many other forms of redundant components are now included in contingency planning, but the ultimate insurance is the provision of either a spare satellite on the ground that can be launched on demand, usually within two months time or so.

Quite large satellite systems such as Intelsat, SES Global or Eutelsat have sufficient satellites in orbit (up to 50 satellites) that they have actual in-orbit spares that can allow a hot transfer of traffic within a few moments of a complete satellite failure. There is elaborate contingency planning to provide for rapid restoral of service, but this is not always possible even if there is an in-orbit spare. In many cases the spare is not a "carbon copy" of the failed satellite. If the spare satellite has a different frequency plan and a different spot beam configuration it may take days to reconfigure the backup satellite to restore most if not all of the traffic. Even if the in-orbit spare satellite is of the same generation and has similar spot beams, there may be failed transponders or problems with the onboard switch to prevent beam interconnectivity.[8]

Satellites are not only configured to restore service in the case of a failed operational satellite but also to restore terrestrial service in the case of a failed terrestrial communications link – especially fiber optic submarine cables. There are so-called

---

[8] *Ibid*, pp. 475–480.

mutual aid working groups that meet to plan for rapid restoration of satellite service by fiber optic links and vice versa. Satellites actually are able to achieve amazing reliability, with availability rates often at 99.99%. The combination of satellites and fiber optic cables with carefully coordinated rapid switchover can achieve availability rates as high as 99.999%.

Submarine cable service can be interrupted by trawlers that dig up buried systems, power outages, etc. Actual outage percentages are higher for submarine cables than for international in-orbit satellites. There is a serious problem with the use of communications satellites for restoration of fiber optic cable systems. This is because many corporate communications systems are designed by chief information officers to be optimized for Internet Protocol (IP) transmissions and with little tolerance for transmission delay. In some instances the network connections are set so as not to allow for routing through a satellite with no allowance for the latency associated with a geosynchronous satellite link. It is therefore imperative for corporate Chief Information Officers (CIOs) to allow for complete flexibility to restore failed telecommunications services, even if this requires adjustment to service standards for transmission delay or issues related to satellite transmission involving Virtual Private Networks (VPNs) and special adjustments that need to be made to cope with IP Security (IP Sec) on GEO satellite connections.

End of Life

Various types of satellite systems in different types of orbit have different lifetimes and different procedures for the disposition of satellites once the satellite is decommissioned from service. Thus, the disposition procedures for GEO systems, MEO constellations and LEO constellations will be addressed separately.

Geosynchronous Satellites

Geosynchronous satellites are generally designed for quite long lifetimes. Lifetimes in the range of 12 to 18 years are quite normal. Lifetime in the case of geosynchronous satellites is heavily based on "expendables" such as fuel for thrusters, the lifetime of batteries and the deteriorating performance of solar cells. There can be a catastrophic failure of a three-axis body stabilization system, power converters, the thermal control system, or other critical subsystem or, of course, a launch failure. If these catastrophic failures do not occur then the mean time to failure of a spacecraft can be reasonably projected to achieve a quite extended lifetime.[9]

In the case of GEO satellite systems neither gravitational effects nor atmospheric drag enter into the calculations, as is the case with lower altitude spacecraft. One can also allow some degree of "inclined orbit" operation toward the end of life of a

---

[9] *Op cit.* Mark Williamson. Pp. 243–266

geosynchronous satellite so that the spacecraft moves north and south off of the equatorial plane to conserve fuel for thrusters. Since over ten times as much fuel is consumed to keep the satellite in the equatorial plane (i.e., north-south station keeping) than the east-west station keeping, the lifetime of the satellite can be extended if the batteries and solar power arrays can sustain operations.

Ultimately, there is a question of disposal of the satellite at the end of life. For geosynchronous satellites that are positioned nearly one tenth of the way to the Moon, returning the satellite to Earth's atmosphere and burning up on re-entry is not a viable option. The pull of gravity in this orbit is 50 times weaker than at sea level on Earth (i.e., 0.22 m/sec2 versus 9.8 m/sec2), and thus it would take the equivalent of a rocket launch to de-orbit a GEO satellite.

The solution is to use fuel to push the satellite outward into a super synchronous orbit, where it will remain in "cold storage" for millions of years. In some cases, however, this push to super synchronous orbit is not possible at the end of life. This might be because the fuel is completely exhausted, thrusters stop functioning or the electronic command capability for the GEO satellite is lost. In this case the satellite drifts to one of two points in the GEO about 300 km up that are both nicknamed the GEO "graveyard orbit." These "graveyard" points (or more technically, the stable equilibrium points) are around 105° W and 285° W.[10]

MEO Constellations

MEO satellites are usually configured in constellations of perhaps 8 to 18 satellites to provide total coverage of Earth's surface. The number, of course, depends of the altitude of the orbit. The higher the altitude of the orbit the fewer the satellites that are needed. Satellites deployed in this orbit have relatively long life because atmospheric drag is virtually non-existent and the gravitation effects are still quite modest. Thus lifetimes of 12 to 15 years are quite common. Since there are a number of satellites in the constellation, it is possible to replenish the network when a satellite fails. Thus one can keep the overall constellation running for quite extended periods of time. The problem comes when a satellite reaches its end of life. A medium Earth orbit satellite is "in between" in many ways. It is lower than a GEO but higher than a LEO. It has less path loss and transmission delay than a GEO but more than a LEO.

This "compromise on performance" is often advantageous, but when it comes to end of life, MEO satellites have a big disadvantage. A MEO satellite is not high enough to be effectively lifted to a super synchronous orbit and it is too high to allow it to easily come down and burn up in Earth's atmosphere. There are guidelines in place for disposal of satellites at the end of life to reduce orbital debris. In the case of the MEO satellite a full 40% of the fuel needed for effective station-keeping must be reserved for de-orbit at the end of life. This requirement constitutes one of the major disadvantages of the MEO. In the early days operators might have

---

[10] *Op cit.* Gary D. Gordon and Walter Morgan p. 75, p. 299, p. 307, p. 325, pp. 371–374.

tended to try to avoid these guidelines for de-orbit and use up their fuel for managing the constellation's operation, but in today's world these guidelines are being taken much more seriously.

LEO Constellations

The deployment of a LEO constellation to provide real-time communications to the entire globe is the most challenging configuration for a worldwide satellite communications network. The large number of satellites in a LEO constellation requires extending the lifetime for the entire constellation. This is primarily because of gravitational pull and atmospheric drag. Although a LEO constellation has the advantage of low path loss (i.e., the least spreading of satellite beams) and minimal transmission delay, there is on the minus side the need to keep perhaps 48 to 70 satellites in continuous operation to complete the network without any "holes" in coverage.

At a low altitude of 500 to 1,200 km there are definite problems of gravitational effects and atmospheric drag that tend to pull satellites to a lower orbit and eventually to de-orbit. This is why LEO satellites tend to have shorter lifetimes, perhaps typically in the range of 5 to 10 years. The low Earth orbit allows for the easier disposal of satellites in that it usually requires only a minimum amount of thrust to start the satellite in an orbit that will lead to its descent into Earth's atmosphere and then burning up during re-entry. The large amount of space debris in LEO orbit and the quite large number of satellites in much closer proximity to one another creates problems with reliable operation and safe de-orbiting.

Although a crash between two satellites (such as the Iridium and Kosmos collision) is rare, impact by smaller elements of space debris is a hazard, particularly to solar arrays. In the case of the Iridium system, which deployed over 80 satellites to operate its network of 66 satellites, for instance, there were concerns expressed by U. S. military satellite operators that the de-orbiting process did not interfere with or collide with "classified surveillance" military satellites. Managing the de-orbit of a large number of LEO satellites is not only an operational issue and concern, but it also raises additional liability concerns as well.

Conclusion

The design, engineering and operation of a satellite system are very challenging tasks. By the end of the 1960s the feasibility of designing and launching GEO communication satellites with high gain antennas that could be continuously pointed toward Earth had been firmly established. The superiority of this design for communications satellites was widely acknowledged throughout the 1970s and 1980s. In the 1990s, however, there were new thoughts about designing satellite systems that would have low latency (or transmission delay) and have much lower path loss. These thoughts to a degree paralleled the emergence of the Internet that had been designed on the basis of rapid network connections with minimal delays. Various attempts were made to come up with new orbital configurations. Some of the new ideas envisioned using LEO and MEO constellations instead of GEO – especially for mobile communication satellite networks that would operate with handheld devices. These new constellation designs, in part, stimulated the idea of finding new ways to construct a large number of satellites with techniques more often associated with mass manufacturing.

Currently, the pendulum has swung back strongly in favor of GEO satellite networks. There is today much more emphasis on designing truly high performance GEO communication satellites with new technology and capabilities and with very large aperture space antennas capable of supporting very small and low cost user antennas on the ground, in the air or on the sea. GEO communication satellites are easier to operate, easier to launch, easier to dispose of at end of life, and generally lower in cost. The engineering challenges continue, and much technology remains to be developed. The context of satellite systems today is that these systems complement terrestrial communications and particularly fiber optic networks and terrestrial wireless systems that carry the bulk of the entire world's communications. Satellites still serve important functions in the areas of broadcasting services, mobile communications and telecommunications connections to rural, remote and isolated areas of the world. Satellites are also important to providing large scale networking services. It is always important to note that market trends and demonstrated service needs tend to drive the development of new technology and IT and communications services. Since there are many existing and new and emergency services that depend on communication satellites around the world, it seems likely that technology in this field will continue to develop and evolve over time.

Earth Stations, Antennas and User Devices

5

Introduction

Satellite engineers focus heavily on the design, launch, and safe operation of space-craft. To satellite operators, the reliable deployment and uninterrupted provision of satellite signals and the successful maintenance of their TTC&M network and spacecraft control center represent their prime objectives. The truth of the matter, however, is that without Earth stations, antennas and various types of user transceivers and user devices, the satellite assets would be worthless. The entire goal of satellite systems design over the past half century has been to find ways to make the ground antennas lower in cost, smaller in size, more transportable and much easier to use. This is in part because Earth station costs dominate the total system costs in most satellite networks.

One of the miracles of the satellite communications industry has been the transition away from huge 30-m parabolic ground systems that were initially used by the first global satellite system Intelsat starting in the 1960s. These gigantic dishes once cost well over $10 million and required a staff of perhaps 60 people to operate on a full-time 24/7 basis. Today there are personal satellite handsets that weigh a few hundred grams and are compact enough that an individual can carry them around and operate them entirely on their own without technical assistance once they have been instructed in their use. This "technology inversion" has allowed us to move from small, low power satellites and huge Earth stations to today's reality of high powered satellites operating off of millions of small dishes and personal satellite handsets and represents one of the major technological achievements of our time.[1]

Today there is a wide array of satellite user antennas on the ground, on the seas and in the air. There are still larger antennas for very high volumes of broadband traffic and for TTC&M operations, but two-way hand-held user antennas and receive-only terminals for receiving radio and television broadcast services have grown into

---

[1] Mark Williamson, *The Communications Satellite*, (1990) Adam Hilger, Bristol, England, UK. Pp. 196–219.

the millions. In general, satellites have gotten larger and more powerful and user devices on the ground have shrunk in size and cost. These user devices vary based on the type of service provided, the frequency band that is utilized and whether the user antenna is for fixed or mobile services, or for receive-only or two way service.[2]

Earth Stations and Ground Antennas

The ground antennas for transmitting and receiving signals to and from the satellite are in many ways similar to the antennas on the satellite. There are several factors that are different that are important to note. One important difference is that signals that come down from the satellite must overcome the noise represented by Earth. Earth, with its solar heat trapped within its atmosphere, is actually "hot" when compared to the coldness of outer space. This "heat" represents a source of interference to the radio wave signals being sent down to the ground from a satellite. The reverse situation does not apply in that the signals from a ground antenna are being sent outward in the coldness of outer space. Thus in terms of receiver performance it is easier to send a radio wave signal out from Earth than to return a signal back to the ground. Further there is an additional problem of actual radio frequency interference that is experienced, particularly in populated areas. This can be interference from nearby satellite Earth stations, terrestrial microwave relay transmissions or even microwave ovens.

Uplink signals have much easier time of it. There is quite simply less interference and noise that must be compensated for since outer space is very "quiet" except for the very, very low noise left over from the "Big Bang." Secondly, ground antennas are able to draw on a large amount of power from national power grids to support their operation. In contrast the satellite antennas are limited to the power that comes from onboard solar cells and batteries. It is because of the limited onboard power that the allocated uplinks for FSS (i.e., 6 GHz, 14 GHz, 30 GHz) are typically at the higher frequencies since these frequencies require more power due to greater "path loss." The corresponding commercial bands for downlinks (i.e., 4GHz, 12GHz, and 20GHz) are thus assigned to the lower frequencies. The same logic applies to mobile satellite services and to spectrum assigned to military communications satellite services, etc.[3]

Antennas for Fixed Satellite Services

The ground antennas for fixed satellite services to a GEO satellite have one key feature that is quite important in minimizing interference. This is simply that the ground station can be located with a clear view to the satellite without any physical

[2] Ibid.

[3] Wilbur L. Pritchard, Henri G. Suyderhoud, Robert A. Nelson, Satellite Communications Systems Engineering, (1993) 2nd Edition, Prentice Hall Inc., Englewood, New Jersey, USA. Pp, 260–262.

barrier blocking the transmission or reception path in any way. Further one can flexibly locate the ground station so as to minimize terrestrial RF interference. In the case of a large ground station for broadband communications and "trunk" traffic one can locate such a facility in a very low noise environment and run a cable connection from the ground station to the terrestrial telecommunications network. For a very small aperture antenna (VSAA) one can still locate the dish in a protected place where RF interference is at a minimum.[4]

In the case of a mobile satellite connection, however, one can at any time experience physical blockage due to trees, tunnels, utility poles or buildings, and there is no special protection from RF interference. Even in the case of BSS services there can be partial blockages from trees or tall buildings at many locations, and there is limited flexibility as to where one places receiving terminals in residential areas. There have been various plans to create fixed satellite services using LEO and/or MEO constellations, but no such system has yet been deployed – although the MEO constellation for the O3b system described below is anticipated in coming years.

The Teledesic system was conceived as a so-called mega-LEO constellation with a very large number of satellites. The unique aspect of this system design was the concept of virtually "painting" static footprints on Earth that would be permanently illuminated by satellites. In short the satellite would be constantly re-orienting its antenna beams so that Earth stations on the ground would not need to track the satellites as they move overhead. To the ground antenna the beam pattern on the ground would seem to always be constant.

More recently there has been the concept of the O3b (Other Three Billion) MEO satellite system. The experience of an MEO constellation such as O3b to support broadband Internet Protocol (IP) services in developing countries could add a new dimension to the type of FSS offerings available, but it would necessitate a new type of ground systems architecture.

Just as is the case with space antennas, ground antennas can be of many types. There can thus be omni dipole aerial antennas, yagis, spiral shaped antennas, horns, mesh parabolic dishes, solid parabolic reflector dishes and phased array antennas.[5] Today parabolic dishes are by far the most common ground antenna for fixed satellite service (FSS) and broadcast satellite service (BSS). There are actually many ways to design feed systems for just the parabolic dish that provide different levels of efficiency. Fig. 5.1 shows a number of different configurations that can be used for a parabolic feed to dish reflector.

The range of ground antennas to support Fixed Satellite Services varies widely in size, cost and throughput performance. The key variables are as follows:
- Frequency Bands. As noted in the earlier chapters the gain of a ground antenna is proportional to the square of the frequency. This can also be explained in terms of the gain of the ground antenna varying inversely with the square of the wavelength.

[4] Mark R. Chartrand, *Satellite Communications for the Non Specialist* (2004) SPIE, Bellingham, Washington, USA. Pp. 261–270.
[5] *Ibid.*

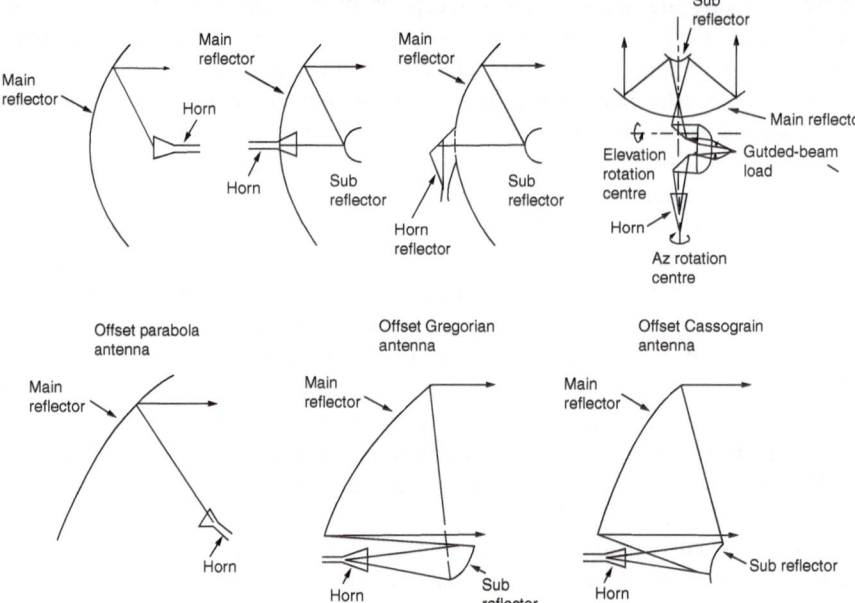

Fig. 5.1 Seven different feed configurations for a parabolic reflector. (Graphic supplied by the author.)

For this reason the C-band antennas usually have wider apertures than Ku-band, and this is even more the case for Ka-band systems. Although the reflectors for the higher frequency services can be smaller (since it is capturing more and more of the tiny wavelengths) the precise shaping of the contours on the parabolic dishes becomes more and more important. This means that the very finely contoured Ka-band ground antenna will be much smaller than a C-band ground antenna if it has exactly the same gain. This does not mean that the Ka-band antenna will not necessarily cost less – and in fact may cost more. If you think of an Earth station antenna reflector as a turned over umbrella and the various wavelengths represented as the size of the raindrops, it is clear that the overturned umbrella can catch a lot more of the smaller raindrops. If you recognize that the "job" of the reflector is to get the incoming radio waves to bounce exactly to the same point where the antenna feed is located then it is also clear that the reflector will have to be more exactly shaped to get the very small wavelengths to bounce exactly where they are supposed to go. And the cost of the ground antenna is not only determined by the cost to manufacture the reflector but also by the cost of the electronics. The electronics that work with the much higher frequencies (and thus much smaller wavelengths) is also more expensive.

• Data Rate or Throughput Requirements. The higher the data rate or throughput requirement for interactive ground stations the higher the gain that is required (and thus the larger the aperture size). A low data rate can be received by a small

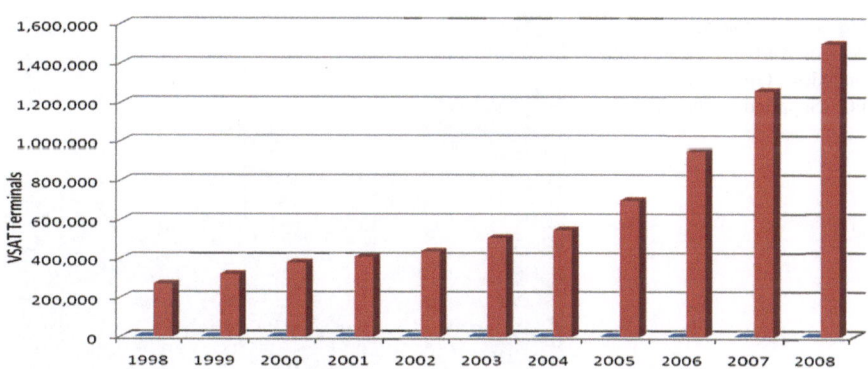

Fig. 5.2 The worldwide growth of VSAT antennas. (Graphic supplied by the author.)

dish, but a high data rate will require a larger ground antenna. (For example a 1-m Ka-band interactive VSAT antenna might typically support an uplink rate of 2 megabits/sec to a FSS satellite, while one might require a 5-m dish to support a 500 megabit/sec link in the same band.) In short, the key aspects in determining the size of a ground antenna are the frequency band that is being used and the throughput requirements.

- Rain Attenuation. A third factor can be that of rain attenuation. If the dish is in an arid climate it can be smaller than in a climate where there are heavy rains and more link margin is required during a heavy rainstorm. This is not much of a factor for antennas operating in the C-band or below, but this must be considered for services provided in the Ku-band and most especially for the frequencies of the Ka-band. This is because the raindrops can work much more effectively as a "lens" to distort and throw off the path of tiny RF wavelengths, especially when the size of the raindrops and the size of the RF wavelength come closer and closer to one another. A raindrop is not likely to distort a radio wave several meters in wavelength, but it can certainly do so to one that is a centimeter or so in length.

The size and cost of FSS terminals continues to decline as economies of scale in manufacturing increase and application specific integrated circuits (ASIC) become more capable, smaller and lower in cost as well. The cost of manufacturing very small aperture antenna terminals (that are often called VSATs) has decreased each year as production volumes have increased. The commercial applications for VSAT networks have continued to increase, particularly with large commercial corporations. These organizations often use VSATs as the mainstay of their global "enterprise networks," such as to verify credit card transactions at ATMs and gasoline stations. This growth can be seen in Figure 5.2.

Today it is even possible to have, at least in the Ka-band frequencies, a "desktop dish" to support 1 megabit/second type interconnections via satellite. The quite compact unit pictured below in Fig. 5.3 has a reflector size that is only 34 cm in size.

Fig. 5.3 Prototype of a
desktop Ka-band antenna that
Includes a GPS and notebook
PC. (Supplied by the author)

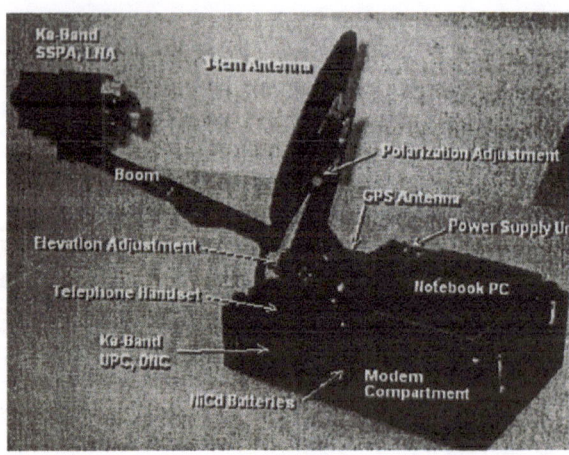

Antennas for Broadcast Satellite Services

The dishes used for most broadcast satellite services, such as for just receiving television and radio programs, are quite simple in design. These types of dishes are often called terminals since they receive the signal but do not uplink a return signal. There are now millions of these types of DBS terminals in operation all over the world. The size of a DBS dish might vary from 1 m in size down to 35 cm in diameter. The size of these dishes will tend to vary based on the transmit power of the satellite, rain patterns in the area and the shape of the transmit beams.[6] In the case of Japan the satellite beams can be very tightly formed and thus allow 35 cm dishes. In comparison a geographically large country such as Australia or the United States will likely have broader beams to cover wide regions and thus there will be a need for dishes that are more like 1 m in diameter to have higher gain.[7]

One of the most important developments in ground antennas operating to broadcast and fixed satellites has been that of Digital Video Broadcasting with a Return Channel Service (DVB-RCS) or a converted service from the cable television world called DOCSIS. Today satellites can be used for digital television or for data distribution with a thin narrowband return data link. The two standards for this type of service are DVB-RCS, as originally developed by the European Space Agency, or Digital Over Cable Service Interface Specifications (DOCSIS), which was originally developed for cable television service but has been found to be quite effective in the satellite world as well.

---

[6] Dennis S. Roddy, *Satellite Communications*, (2001) 3rd Edition, McGraw Hill, New York, USA pp. 209–215.

[7] Joseph N. Pelton, Satellite Communications 2001: The Transition to Mass Consumer Markets, Technologies and Systems, (2000) International Engineering Consortium, Chicago, Illinois, USA. pp. 93–94.

Since there was more of the equipment initially developed for cable television DOCSIS-based networks, these are sometimes lower in cost. This type of service can operate on either Fixed Satellite Systems or Broadcast Satellite Systems and is very efficient for large corporate networks. The digital broadcast service may operate at high speeds of 36 megabits/sec, 45 megabits/sec or even 72 megabits/sec and with a return link that is anywhere from 64 kilobits/sec upward. Large corporations that wish to distribute sales information, training videos, and advertising videos and receive in return from local or regional offices inventory updates or encrypted e-mail messaging have found this type of service, sometimes referred to as moving service to the "edge," as very effective. This means there is an ability to provide very high data rate service to highly distributed local users. Such digital broadcast service with a return capability is used by broadcasters, large multi-national enterprises and even for military applications.[8]

Antennas for Mobile Satellite Services

The greatest challenge in designing ground antenna systems for satellite systems is undoubtedly that of mobile satellite services. This is because of the triple challenge to be overcome.[9]

- Challenge # 1. Need to supply MSS users with a highly mobile, compact and reasonably low cost hand-held unit that seems reasonably comparable to a cell phone.
- Challenge # 2. Need to have sufficient margin so that one can operate the satellite phone without the circuit being broken when driving through a forest, or near a building or other obstruction, and be able sustain a call from an automobile.
- Challenge #3. Need to make sure that interference from other handsets is manageable if there are several units operating in proximity to a satellite telephone.

These were technical challenges that the Iridium and Globalstar mobile satellite systems first had to face, and this posed many difficulties for them. Some very imaginative engineering was employed to try to meet the above challenges. Some of the technical solutions employed involved having a magnetically attached antenna that could go on top of a car, extendable antennas that could be pulled out to increase gain and spiral antennas hidden within a tube covering. These higher gain antennas would receive overhead signals but not from an underground location. The challenge of making these satellite handsets small enough to be convenient, but not making their antennas so small that they could not sustain a conversation, was tremendously demanding. Over time more and more capable application-specific integrated circuits (ASIC) have allowed improvements in performance.

[8] Joseph N. Pelton, *Future Trends and Satellite Communications Markets* (2005) International Engineering Consortium, Chicago, Illinois, USA pp. 33–34.
[9] Gary D. Gordon and Walter L. Morgan, *Principles of Communications Satellites* (1993) John Wiley and Sons, New York, USA. pp. 46–47 and p. 90.

Other mobile satellite systems deployed after Iridium and Globalstar abandoned the LEO constellation approach. These later systems have sought to use GEO satellite technology and have designed huge deployable (or unfurlable) wire mesh antennas up to 22 m in diameter. The technical concept of these later mobile satellite systems is to create very tight spot beams off of these huge reflectors and then be able to interconnect these beams. These spot beams on the satellite generate enough power to allow the handset and their antennas to be quite small.

An even newer concept is to make these systems "hybrid" so that the handsets are "dual mode." This means that the handset will operate as a terrestrial cell phone within a city or on an interstate highway, but will automatically switch to a satellite connection in a rural area without coverage. The question remains as to whether the consumer will find the handsets that work with systems such as Inmarsat (especially *Inmarsat 4*), Thuraya and now Terrestar and Light Squared low enough in cost, easy enough to use and reliable enough in service to make these systems a commercial success. (See Fig. 5.4.) The initial financial difficulties associated with the early land mobile satellite systems of Iridium and Globalstar have led to many market analysts expressing concerns.

Fortunately, the mobile satellite service to meet the needs of maritime communications is now well established, and this service has steadily evolved to provide reliable and extensive aeronautical mobile satellite service as well. Maritime and aeronautical satellite service enjoys one advantage over land mobile satellite service. Ships and boats on the high seas and aircraft in the air enjoy a direct and unobstructed view of the satellite without having to worry about forests, billboards, buildings or tunnels blocking the pathway to the satellite. Inmarsat has been providing maritime service since the 1980s, and other providers such as Thuraya have supplemented this type of service. The user antennas on ships and aircraft have become smaller and more user friendly without sacrificing performance. The initial shipboard antennas were on the order of 3 meters in size and had to be protected by

Fig. 5.5 Compact broadband
ship-mounted antenna for
Inmarsat 4 operations.
(Graphic by permission from
Inmarsat.)

a plastic conic shield and radomes. They also had to have a complex tracking and stabilizing system to orient the high gain antenna so that it could be constantly pointed to GEO satellites.

Today, with the Inmarsat 4 high gain satellite in space with its 12-m antenna, ships can communicate at speeds such as at 500 kilobits via small ship-mounted antennas such as pictured in Figure 5.5. Passengers can even use handheld phones for communications via Thuraya, Inmarsat, Iridium or Globalstar to communicate at much lower speeds. Unfortunately these handsets are not multi-purpose and thus a particular handset only works with one particular satellite network. Conformal antennas shaped to the sides of aircraft or helicopters can likewise communicate to Inmarsat or Thuraya satellites in GEO or to Iridium or Globalstar satellites in LEO.[10]

The bottom line is that there is no place on Earth's surface, or in the oceans or in the skies, that cannot be reached via a satellite transmission today. People can connect with satellite hand phones to a variety of different satellite systems via voice, video or data. These satellite phones are still proprietary in nature, so one must purchase or lease a particular handset to work with a particular satellite. There is a wide range of commercial and defense-related communications that mobile satellite services support today. These mobile satellite connections have also proven to be life savers for explorers who undertake risky adventures such as climbing mountain peaks or flying in balloons or sailing solo across the oceans.[11]

---

[10] D.K. Sachdev and Dan Swearingen, "The New Satellite Services: Broadcast, Mobile and broadband Internet Satellite Systems" in Joseph N. Pelton, Robert J. Oslund and Peter Marshall (editors), Communications Satellite: Global Change Agents (2004) Lawrence Erlbaum Associates, Mahwah, New Jersey, USA, pp. 69–81.

[11] Ibid, pp 74-80.

Antennas for Store and Forward (Machine to Machine Service)

There are a variety of simple antennas used to support store and forward satellite systems such as Orbcomm as well as a number of "Surrey sats" (satellites from the Surrey Space Center in the UK) that support a variety of national and international missions. These satellites, as well as the OSCAR satellites built and launched on behalf of the amateur satellite organization, are low in power and are usually able to support only data communications. The only truly unique feature of these store and forward satellites in terms of ground systems has been the convenient combination within the Orbcomm network of both data satellite communications services and GPS space navigation capability. This type of hybrid data and position location service has been effectively used by shipping lines, trucking services and car rental networks. These industries have found it to be quite attractive to have both two-way texting communications plus an active space navigation capability that shares a common battery all within one low cost transportable unit.

Antennas for Tracking, Telemetry, Command and Monitoring

One of the key elements of a satellite network is the tracking, telemetry, and command system that provides near real-time information about the health of the satellite and allows tracking and control of the satellite. Larger and more capable Earth stations, often operating at different frequency bands, perform these critical functions. Many operational satellite systems also undertake an active monitoring system that checks on the quality of the satellite communications services being provided on the system and seeks to detect frequency interference or other problems with the operational transmissions.

These facilities are often co-located with operational Earth stations in the case of GEO orbital systems, but in the case of LEO and MEO systems a much larger TT&C network is required because of the need to have visibility of a large-scale constellation at locations all over the world – unless there are inter-satellite links onboard the satellites in the constellation, as is the case with the Iridium network. In the case of Globalstar scores of TT&C facilities had to be put in place. In order to maintain instantaneous contact with all satellites in this 48-satellite network, plus its spares, a network of some 70 stations would have been necessary. Small gaps in coverage have been considered acceptable, and thus a much smaller TT&C network has been used by Globalstar. These TT&C ground stations, as well as the monitoring stations, typically use conventional parabolic dishes and thus require no special design or engineering. Such networks are particularly critical during launch and test operations.

The Future

Ground stations are a critical part of a communications satellite system. There continues to be remarkable growth in the technology that allows satellite ground antennas to become more capable while being manufactured at lower and lower costs.

In some ways the fabrication and delivery of VSATs has become comparable to the manufacture and sales of automobiles in a global market in terms of efficiency of design, production and global distribution. The process of technology inversion has traveled a long way so that increasingly sophisticated satellites can support ground antennas that are more versatile, lower in cost, smaller in size, and in some cases more mobile, and easier to install and operate. In short, as in-orbit satellites have become larger and more complex, the greatly increased number of ground antennas have become smaller and easier to use. The drive behind this process has been to decrease overall systems costs. If there are millions of user devices on the ground that can be reduced in cost by only a few hundred dollars per unit this easily justifies spending $50 million or even $100 million more on the space segment.

Until some limit is reached in terms of total systems costs (or another limit is reached such as health concerns with regard to operating mobile units that radiate power close to the human body), this process will likely continue. There are certainly smaller satellites (known as micro satellites or nano satellites) that are operated by countries or organizations that cannot deploy satellites costing hundreds of millions of dollars, but these satellites are not as cost effective in terms of total systems expenditures and derived capacity and performance. Also such small satellites contribute significantly to the growing problem of space debris.

Critical technologies for ground systems are application-specific integrated circuits and improved batteries. Both of these technologies have improved greatly in recent years. In many cases the greatest barrier to the further development of satellite services around the world may well be the tariffs that are applied in many countries to small satellite ground antenna systems. In some cases import tariffs for a mobile satellite handset or a receiver for direct audio broadcast service or some other ground antenna, particularly in developing countries, may serve to double – or more – the cost of the actual device. In short trade, tariffs and regulatory restrictions on satellite ground antennas may pose the largest barrier to the future growth of satellite services – far more than technology or operational constraints.

Satellite Deployment, Station-Keeping and Related Insurance Coverage

6

Introduction

Without reliable launch services there would be no communications satellite industry. The good news is that the reliability of satellite launchers has continued to increase over time. The bad news is that despite high expectations launch services have not really gotten cheaper. The most important development in launch services over the last half century is probably the fact that a number of countries and launch services organizations within those countries have developed launch capabilities to low, medium and geosynchronous orbits. This competition has helped to lower costs. Also the development of larger boosters has allowed the launch of larger and more cost-efficient satellites (in terms of total systems costs). However, the scale factor that has allowed many products to be more cost effective over time has had limited impact on reducing launch costs.[1]

A key development that has allowed a wider range of organizations to enter the field of satellite communications worldwide has been the development of launch insurance. Initially such coverage was almost impossible to obtain and was quite expensive as well. Today this industry is well established, and a global process of reinsurance has allowed insurance companies to avoid having a wide exposure to a single launch or in-orbit satellite failure.

Chemically powered launch vehicles and thrusters are today the mainstay technology for launching satellites and maintaining them in orbit, but there are new technologies on the horizon that might be used to deploy satellites and maintain them in orbit at lower cost, with greater reliability and perhaps less adverse environmental effects. Future systems that might be available in future years will be addressed at the end of this chapter.[2]

---

[1] Joseph N. Pelton, *Basics of Satellite Communications*, (2006) International Engineering Consortium, Chicago, Illinois, pp. 142–147.

[2] Mark Williamson, *The Communications Satellite* (1990) Adam Hilger, Bristol, England, UK.

J.N. Pelton, *Satellite Communications*, SpringerBriefs in Space Development, DOI 10.1007/978-1-4614-1994-5_6, © Joseph N. Pelton 2012

Key Developments in the Launch Services Industry

Perhaps the most important new factor in the global launch service industry has been the recent evolution of new, highly entrepreneurial companies that are seeking to develop a wide range of new space-related products and services. This activity had its "explosive start" with the organization of the "X Prize," which was initially aimed at trying to start a commercial space tourism market. But today, it has morphed into a commercial space transportation business. The initial X Prize was won by Burt Rutan of Scaled Composites with the financial backing of Paul Allen, the co-founder of Microsoft. Most recently multi-million-dollar Google Lunar X Prize has spurred a wide range of innovation in the launch industry. Likewise the NASA Commercial Orbital Transportation Systems (COTS) program has spurred the development of new launch capabilities by Space X (The Space eXploration Corporation), Orbital Sciences and others.[3]

It is felt by some that the development of totally new launch capabilities that transcend chemical rocket technology is the only real way to develop reliable and low cost access to space over the longer run. It seems likely that the next twenty years will see chemical rocket boosters remain as the core technology to getting communications satellites placed into orbit. In the longer run, many believe that safer, easier, greener, less costly and more reliable ways will be found to get satellites and people into space. There are certainly concepts about how this might be accomplished such as electrical (ion propulsion), nuclear propulsion, rail-guns, tether-based systems and perhaps ultimately a space elevator. To date commercial communications satellite operators must continue to rely on chemical rockets and about a dozen suppliers of launch services worldwide.[4] (See Table 6.1 below.)

Today the conventional way to orbit is by using chemically powered launches that boost a satellite into orbit. For launches into geosynchronous orbit, the most typical arrangement is to launch a satellite into a so-called transfer orbit. This is very high elliptical orbit with the apogee being essentially at the height of a geosynchronous orbit. On a particular orbit where the apogee is near the desired point on the geosynchronous orbit, a rocket propels the satellite from the transfer orbit into a perfectly circular orbit along the equatorial plane. In past decades, when there were only a few satellites in GEO, this was not such a difficult and delicate operation as it is today. Today there are over 400 active or inert satellites in GEO and the number grows each year. Today this must be a precise operation to not risk interference with other satellites in GEO.

Today the European Arianespace organization, based in France and at the Kourou, French Guiana, launch site, dominates the commercial communications satellite market by providing the majority of launches. The Ariane 5 is capable of launching the world's largest communications satellites or launching two or more

---

[3] Joseph N. Pelton and Peter Marshall, *Space Exploration and Astronaut Safety*, (2006) AIAA, Reston, Virginia, USA, p. 277.
[4] *Op cit*, Pelton, p. 145–147.

Table 6.1 Global review of launch service providers and their launch sites. (Supplied by the author.)

GLOBAL LAUNCH PROVIDERS FOR COMMUNICATION SATELLITES		
Launch System	Launch Service Provider	Most Frequent Launch Site
Ariane 5 Launchers	Arianespace of Toulouse, France	Kourou, French Guiana (CSG)
Delta Launchers	Boeing Corporation	Cape Canaveral, Florida, and Vandenberg, California, USA
Atlas V Launchers	International Launch Services	Cape Canaveral, Florida, and Vandenberg, California, USA
Pegasus Launcher	Orbital Sciences	Launched from Converted Lockheed L1011 aircraft
Taurus and Minotaur Launchers	Orbital Sciences	Wallops Island Flight Facility Vandenberg, California, USA
Falcon 1 & Falcon 9	SpaceX Corporation	Cape Canaveral, Florida
Long March Series of Vehicles Launchers	Great Wall Corporation	Xichang Launch Center, China
Proton Launchers	International Launch Services and Khrunichev	Baikonur, Kazakhstan
Soyuz	STARSEM (French/ Russian Consortium)	Baikonur Cosmodrome and Korou, French Guiana
Zenit 3SL	Sea Launch	Pacific Ocean at Equator
Rockot 1	Kurichev of Russia	Plesetsk, Russia
H2 and H2A	Japan Rocket System Corporation	Tanegashima Launch Center, Japan
Geo and Polar Satellite Launch Vehicles (GSLV & PSLV)	Indian Space Research Organization (ISRO) Vehicle Commercial Subsidiary	Sriharikota Island, India
Zenit 2	Yuzhnoye, Ukraine	Baikonur, Kazakhstan

Table by Joseph N. Pelton, 2010.

communications satellites at the same time. Indeed the predominant launch capabilities today are of the largest class of launcher such as the Atlas V, the Russian Proton, the Japanese H2A and the Zenit 3. These rockets can launch at least six metric tons into GEO, and the Ariane 5 can launch an amazing 10 tons to GEO transfer orbit. These rockets have tried to achieve increased cost efficiency by using scale to their advantages.[5]

There are nevertheless a wide range of launch options available for different classes of satellites. There are a range of launchers suited to a variety of satellites

---

[5] Mark Chartrand, *Satellite Communications and the Non Specialist*, (2004) SPIE, Bellingham, Washington, USA, pp. 185–198.

Fig. 6.1 The Falcon 9 launch vehicle illuminated at night. (Graphic courtesy of Space X.)

and different orbits such as the Pegasus, Taurus, Minotaur, Delta, Falcon 1, Zenit 2, Rockot 1 and Indian PSLV vehicles able to accommodate launches of smaller space-craft and to different orbits. One of the newest and potentially most exciting options is the new Falcon 9 series of launch systems that Space X is developing. This is an entirely new launch vehicle that is being developed by Elon Musk, who made over a billion dollars from his sale of PayPal, the on-line secure payment system. The Falcon 1 and Falcon 9, in contrast to the great bulk of available launchers, represent a highly entrepreneurial and streamlined development enterprise.

Most launch systems listed in Table 6.1 have been developed as national launcher development programs under the oversight and guidance of either national space or military agencies. These programs have put performance and reliability as the top goals and cost efficiency has not been the top priority.

The successful development of the Falcon 1 and 9 series of launch systems and the Dragon flight system is thus perhaps the first entirely entrepreneurial venture that has given equal weight to cost efficiency and to performance. The objective is to develop launch vehicles that are reliable and suitable for commercial satellite launches but at significantly lower cost than other commercial launchers currently available in the marketplace. The successful first launch in 2010 of the Falcon 9 launch vehicle gives a considerable boost to new commercially developed and higher capacity launch vehicles. The space shuttle is not included in Table 6.1 because this launch vehicle is now out of service. Further these were vehicles used essentially for launches with crew aboard and not for unmanned application satellite launches.[6] (See Fig. 6.1.)

---

[6]"Successful Launch of Falcon 9" www.spacex.com/**falcon9**.php

The obvious factors that a commercial organization looks for in terms of launch services are reliable performance, reasonable cost, and compatibility with launch schedule needs. There is, however, an important fourth factor that cannot be overlooked. This factor is the technical evaluation of the compatibility of the satellite and the launch vehicle.

The starting point is to ensure that the satellite properly fits within the nose fairings of the satellite with sufficient margin. Then one must assess if the physical interface will not give rise to harmful vibrations, harmonics, pogo effects, radiation levels and/or damaging thermal levels during launch. All of these possible harmful effects could disable the satellite during launch or affect the spacecraft so that it does not function properly in orbit.

Alternatively the satellite could be electronically and physically functional after going through launch, but it could turn out that it is not possible to deploy the solar array or the antenna systems, thus rendering the satellite useless. Extensive thermal vacuum tests, vibration tests, and shaker table tests are undertaken to ensure that the satellite can withstand the many potentially disabling effects experienced in a high thrust, multiple g-force satellite launch. It is often the case that trial launches are undertaken with the rocket equipped with a number of onboard sensors to determine and record the many potentially disabling factors that could potentially disable the satellite during launch or prevent one or another of the elements of the satellite to be properly deployed. There are particular "harmonic frequencies" that need to be avoided that could shake the satellite so that its electronics or other components malfunction.[7]

Launch Operations

One chooses to launch rockets from specific launch sites for a variety of reasons. A first consideration is safety. This typically means that the site is on a coastline so that one can launch over the ocean, away from inhabited areas. An alternative can be a very rural and remote area such as Siberia, Russia, the Australian Outback, etc. Not only is there a requirement to launch away from populated areas, but there are other safety and environmental considerations. A number of special facilities are needed for storage of explosive materials. There is a need for range safety control that can monitor and abort the launch and also to coordinate with aviation safety control so that a launcher and aircraft have no possibility of colliding.

Perhaps the most important consideration is finding an optimum launch site to provide the best and lowest thrust access to the desired orbit. In the case of a GEO launch it is desirable to have a launch location as close to the equator as possible. Since Earth is rotating at the equator at a speed of about 1,600 km/h, the closer the launch site is to 0 degrees latitude the greater the "assist speed" to the launch.

[7]Gary D. Gordon and Walter L. Morgan, *Principles of Communications Satellites*, (1993) John Wiley and Sons, New York, pp. 448–457.

Fig. 6.2 The sea launch platform for ocean-based launch of the Zenit 3SL vehicle. (Photo courtesy of Sea Launch.)

The Kourou launch site in French Guiana, which is quite near the equator, has a significant advantage over Cape Canaveral, Florida, in terms of liftoff advantage toward GEO. One of the more clever solutions to launching satellites into GEO is that devised by the Sea Launch Corporation with a launch platform that leaves port out of southern California and then launches a modified Zenit launch vehicle from the Pacific Ocean at a position near the equator. Although Sea Launch has had major financial difficulties it has been financially reorganized and plans to restart launch operations during 2011.[8] (See Fig. 6.2 below.)

Countries that are well away from the equator, such as Japan and China, experience a clear disadvantage in terms of having a national launch site suited to GEO launches. Launch operators that are seeking a polar orbit, on the other hand, can benefit from launch sites with a high latitude.

---

[8]"Cruising to Orbit" http://www.sea-launch/sllaunch_vehicle.htm

One of the key challenges for GEO launches is to have a global network of Tracking, Telemetry and Command Earth stations to support a launch. Organizations with such a global capability such as Intelsat often lease their facilities to other organizations to support their launch operations.

The Future of Launch Systems and Station-Keeping

There have been ongoing efforts to make launch systems more reliable and more efficient. In the case of aircraft, 13% to 15% of the vehicle's weight is its payload, and the rest is fuel. This may not sound too impressive, but typical chemically fueled rockets only have 1% to 2% of their weight devoted to lifting a payload to orbit, and the rest of the weight is essentially the rocket fuel, the rocket structure and the rocket engines. Only the powerful Saturn V vehicle was able to dedicate 5% of its weight to payload.[9]

The object in developing future launch systems is to find a way to achieve orbit with much greater efficiency and, of course, greater reliability. Some of the concepts that are under study include the following.

Electrical Propulsion Using Ion Engines for Station-Keeping. These are much lower thrust systems than chemically powered rockets, but they provide thrust for much longer. In terms of mass-to-thrust ratio over time they are more efficient than chemical rockets and can be more cost effective. Ion engines today are used for thrusters for station-keeping and could be used over a period of many weeks to lift a satellite from LEO to MEO or even GEO.

Nuclear Powered Thrusters. Nuclear isotope-based systems to provide electrical power in space, and especially those that can provide electrical power to propel ion engines, have already been extensively tested in space. For safety reasons nuclear propulsion based on either active nuclear fusion or fission processes are not being developed. Nevertheless nuclear-based systems such as the latest versions of the so-called SNAP generators could produce high levels of electrical power sufficient to create ion engines of much greater thrust than today's systems. There are no current plans to use nuclear-powered vehicles to launch satellites from the ground, although there is the idea of using nuclear propulsion for missions in outer space such as exploratory probes within the Solar System.

Tether-Based Lift Systems. Such systems could lift payloads from LEO to much higher orbits at low cost and high reliability. The development of such tether lift systems could be a crucial path forward to developing over time so-called space elevator systems.

---

[9]Op Cit, Williamson pp.267–282.

Space Elevators or Space Funiculars. NASA and entrepreneurial companies such as the Space Elevator Company are currently seeking to develop the enabling technology to make such a lift to orbit system possible. These developments include solar-powered climber robots, carbon based nano-tube systems and so-call "Bucky Ball" materials that could be fabricated into 100,000-km-long cable strands. A number of technologies and materials need to be developed to create a viable lift system between Earth and the Clarke Orbit. Although such a system may be many decades away, it would not only revolutionize the cost of launching GEO communications satellites but also the cost and reliability of all forms of space travel.

Space Insurance and Risk Management

The initial arrangements for so-called launch insurance provided quite limited protection indeed, with a cash settlement for only partial coverage of the total loss and then only after a second consecutive loss. The aviation insurance industry over time developed models based on losses and reinsurance programs that are used to develop insurance policies for aircraft and other high risk projects such as nuclear power plants. The arrangements took into account the history of launch and failure for different launch vehicles, the number of launches that were to be insured, steps that satellite operators took to minimize risks, and the value of the satellites being launched. The true key to the launch insurance for communications satellites was the arrangement through national and regional insurance markets to spread the risk, first through a few firms and then reinsurance arrangements whereby perhaps fifty or more insurance companies take a small share of the overall risk.

Some of the largest satellite communications operators such as Intelsat led the way forward in the launch insurance arrangements. Intelsat sought coverage for an entire satellite series and made arrangements for the insurance for six or more spacecraft launches at a time. It also hired expert staff with particular knowledge of launch vehicle technology to oversee all aspects of a satellite launch in order to minimize the possible mistakes or accidents that might lead to a launch failure. The most successful launch insurance policy ever negotiated at least for a satellite service provider was 7% of the insured value for the satellite and launch vehicle. The typical cost of launch insurance today will likely range from 15% to 20% of the insured value. Some of the very largest organizations, however, will self-insure their satellite launches. Smaller projects with just one or two satellites almost always feel that launch insurance is the best arrangement even if their premiums are at the high end of the market scale.

Today one can insure much more than just the value of a satellite and its launcher. One can negotiate almost any type of coverage desired, such as coverage for successful operation of a satellite in orbit on an annual basis that is renewable. One can insure against lost revenues until a replacement satellite is launched. One can insure against a collision with another satellite.

One of the key risk issues is that of liability if a rocket should go off course and land in a heavily populated city and create enormous damage and billions of dollars in claims. Since the launch operators and range safety officers control explosive devices that can destroy a rocket and abort a mission if it should go off course, the risk of such an event occurring is quite small, but the risk factor of financial exposure is enormous. Different countries with space launch programs and with launches of military and governmental satellite programs have had to face this possibility. Some countries provide for coverage against all liability claims above $500 million and require commercial operators to purchase liability coverage up to $500 million. Other countries require a sharing of the risk. Since the risk of a catastrophic launch with a large number of fatalities is small, liability coverage costs are much lower than the launch insurance for the satellite and the launcher.[10]

One practice that has developed in recent years for large aerospace companies that build communications satellites is for them to offer what they call a "turnkey" package for a fixed price or a fixed price plus incentives. In this type of arrangement the aerospace contractor agrees to provide a number of satellites with guaranteed performance in orbit for perhaps several months. Under this "turnkey" or "delivery in orbit" arrangement, the contractor arranges for launch services, launch and liability insurance coverage, and perhaps the Earth segment equipment as well. Under this arrangement the aerospace company becomes the "general contractor" that makes all of the arrangements and absorbs some degree of risk – depending on how much insurance is obtained and the conditions of this insurance coverage. In such an arrangement, if the satellites are successfully deployed and perform to specs the contractor can make a larger profit. If there are difficulties the contractor may end up making only a very small profit or even lose money.[11]

Conclusions

The development of launch systems has not kept pace with the development of new communications satellite technology or new ground antenna systems in terms of increased cost efficiencies. The innovations in digital technology, monolithic devices and other technical developments have allowed satellite and ground systems to move swiftly ahead and make rapid gains in terms of throughput, cost efficiencies, and general performance. The recent development of the new Falcon 9 vehicle suggests that entrepreneurial talent and new technology may provide a spurt toward new, more cost effective launch vehicles. The key hope for the future is that totally new systems to provide access to orbit might come from technology that is only now beginning to be seriously developed. In short, it is hoped that new launch systems can eventually arise from electric propulsion, ion engines, rail guns, tether lift systems, and even the long dreamed of space elevator.

---

[10] *Ibid* pp. 267–280.
[11] *Ibid* pp. 281.

Key Business, Trade and Regulatory Issues

7

Introduction

The satellite industry is among the most complex of any business in the world. Success depends on developing and implementing new technologies in space and on the ground as well. It also requires implementing the most modern of operating systems that utilize the maximum amount of automation. Perhaps of equal importance are strategic, trade, business and regulatory issues. To operate a successful global satellite business requires the ability to have access to a range of assigned frequencies for satellite communications operating on satellites with internationally registered orbital locations within the international regulatory processes of the International Telecommunication Union (ITU). There is also the need to formally establish "landing rights" in every country where service is provided.

There are also a host of technical standards to be observed from national, regional and international standards organizations plus trade restrictions and processes that derive from the World Trade Organization (WTO) and the General Agreement on Trade in Service (GATS). On top of all these technical, operational, trade, regulatory and standards issues, the satellite industry is one of the most competitive industries in the world. Satellite systems providers must have the latest and most cost effective satellites and ground stations, meet technical standards, and jump through all of the necessary regulatory and procedural hoops just to compete. Then they must be able to win customers against highly competitive satellite service providers as well as terrestrial service providers of fiber optic networks and broadband terrestrial wireless providers.

Regulatory and Frequency Allocation Issues

All operators of satellite communications must start with the design of a satellite system that uses allocated radio frequencies for the services that they wish to provide. They must first get their national governmental agencies to agree to support the licensing of such a satellite network and send the technical characteristics of the satellite (frequencies and orbital location or constellation configuration) to the ITU for international coordination. If any other nation responds to the ITU international posting of this satellite filing and indicates they believe this system will interfere with their own network, then an international coordination process is initiated.

There have been some proposed satellite systems that have sought to circumvent the demanding national procedures of their own governments and gone to other "non space" countries to file their applications with the ITU. These have been called "flag of convenience" filings. There have been some other jurisdictions that have been accused of filing "paper satellites" with the ITU to seek certain benefits from satellite operators. To combat such practices, filings with the ITU now require very substantial filing fees. In addition, jurisdictions that have engaged in such practices, such as the Island Kingdom of Tonga, Gibraltar, the Isle of Man and Jersey, have been discouraged from such end runs by ITU procedural changes that make it more difficult to file "paper satellites" without the signing of contracts and other concrete steps to validate their filings.[1]

The ITU through a process known as a World Radio Conference, a plenipotentiary of all of the ITU members (some 200 in number), gets together and agrees how to allocate spectrum for a wide range of wireless services, including all forms of satellite communications for commercial purposes as well as for governmental and defense purposes.

Global Trade Issues for Satellite Communications Services

The WTO that was formed to take over the activities of the GATT has increasingly shifted its focus from trade in goods to trade in services. The adoption of the GATS covers telecommunications and thus satellite services. The process, however, is a complicated one. Member countries of the WTO and subscribers to the GATS are left to develop a voluntary set of guidelines on how much competition they will allow with regard to telecommunications services, the extent of foreign ownership, and the rules that apply to foreign providers of satellite services. In some countries the guidelines to allow competition in satellite services have been adopted, but in practice global providers of satellite services have often found it difficult to find local partners. In some cases, this is because the local government has quietly indicated that it would rather not see competition to national satellite service providers.

The WTO is increasingly moving to place sanctions on countries that do not allow reciprocal trade in telecommunications and satellite services. As key countries such as India and China come into the WTO framework, together with countries that have been slow to embrace full competition, such as Brazil and South

---

[1] Ram Jakhu, "ITU Regulatory Procedures for Satellite Network Filings." International Space University Lecture.

Korea, it will be increasingly easy to provide worldwide communications satellite services across the world. In some countries the extent of competition depends on whether one wishes to deploy competitive satellite facilities or to compete for satellite services based on leased capacity from national service providers. In the case of Japan the regulatory process defines as a Type 1 carrier an entity that wishes to provide telecommunications or satellite facilities. To become a Type 1 carrier is a long and painstaking process, and only a few Type 1 carriers are allowed under Japanese law. To become a Type 2 carrier based on leased services is a much more widely accessible activity. Although there are requirements to operate a local office and have an engineering staff, becoming a Type 2 carrier is generally open to all organizations.

Although there have been attempts to broaden the ability to compete for communications satellite services around the world, the truth is that there remain many constraints to open competition in many countries. Especially if a company seeks to deploy new satellite systems (as opposed to just competing at the service level on existing networks) the pathway can be filled with many regulatory and standards-related roadblocks.

Technical Standards

(Related to Satellite Delay and Internet Protocol Based Services)

The number of technical standards that apply to providers of communications satellite services is quite broad. These standards are constantly evolving, and service providers must expend a substantial amount of time and money to follow and participate in the standards-making process. There are important national and regional standards-making bodies such as the American National Standards Institute (ANSI), the Inter-American Telecommunications Commission (CITEL) of the Organization of American States, the Asia Pacific Telecommunity (APT) and the European Telecommunications Standards Institute (ETSI).

The ITU has a number of working groups within the Consultative Committee on Telephone and Telegraph (CCITT) that addresses terrestrial telecommunications and numerous other working groups within the Consultative Committee on Radio that address wireless and satellite technical standards. In addition there are standards-making groups that develop interface standards and technical guidelines related to satellite communications all over the world. If you would like more detail a listing of relevant standards-making bodies can be found in the end-notes of this chapter.[2] In many ways these organizations work together, and there are many forms

---

[2]Groups that develop standards that relate to telecommunications and satellite communications also include the International Electro-technical Commission (IEC) in Paris, the International Standards Organization (ISO) in Geneva, the Institution of Electrical and Electronics Engineers (IEEE), the Motion Picture Expert Group (MPEG), the Internet Technical Committee, and the Internet Engineering Task Force (IETF). In addition there are a series of standards that relate to government and defense-related telecommunications systems.

of cooperation that assist with an orderly process to create viable standards for satellite communications. One of the large challenges is that while many of these standards-making groups work largely within the framework of the ITU and traditional U. N.-based standards-making processes for telecommunications systems, there is a largely independent and terrestrially oriented set of processes that develop standards and operating guidelines for the Internet. The Internet Engineering Task Force (IETF) works through a process of sending out proposed operating standards as Requests For Comments (RFCs). This often ends up with a set of operating standards related to the Internet Protocol (IP) and Internet Protocol Security (IPSec) that are not always fully compatible with efficient satellite transmission.[3]

For a number of years satellite engineers have sought to develop what are called Internet Protocol over Satellite (IPoS) standards that allow efficient transmission of Internet-based messaging via satellite. These processes have devised ways to distinguish satellite latency (or satellite transmission delays) from terrestrial system congestion. They have also helped to solve other problems such as ways for satellites to serve virtual private networks (VPNs) effectively and to allow satellites to cope with the special needs of IP Security (IP Sec) operations.

In the early days of the Internet transmission over satellite throughput rates and transmission efficiencies were quite low due to all of these standards issues, and the fact that IP had been optimized for terrestrial networks and not satellites. Today, with the creation of IPoS standards and improved standards-making, efforts aimed at increasing broadband Internet transmission via satellite efficiencies have increased to over 80%.[4]

Competitive Business and Strategic Issues

There have been several key sea changes in the satellite communications world in the past few decades. These major shifts in the industry might even be described as "satellite megatrends." These megatrends include:

- Technology Inversion. The shift in satellite and Earth sector systems that might be called "technology inversion" has seen the satellite world transformed from an industry where small and power-limited satellites operated to large Earth stations. Today the world of satellite communications is one where large and powerful spacecraft now operate to literally millions of small ground and mobile antennas;
- Diversification of Satellite Services. Once the satellite industry was focused almost exclusively on international Fixed Satellite Services (FSS). This is no longer true. Today it is a highly diversified industry with a large number of

---

[3] Mark R. Chartrand, "Satellite Communications for the Non Specialist" (2004) SPIE, Bellingham, Washington, p 261–63.

[4] Hughes Network Systems, "IP over Satellite—The Standard for Broadband over Satellite" www. hughes.com/.../IPoS_H30740.pdf

different markets. These markets now include broadcasting (both television and radio); domestic, regional and international fixed satellite services that support voice and television distribution and a variety of data- and Internet-based services including services to the small office and home office (Soho); a wide diversity of mobile satellite services; and store and forward data distribution and machine to machine communications. Currently international fixed satellite services probably are under 10% of the industry's total revenues.

- The Shift from Monopolies to Highly Competitive Systems. Once satellite communications services were largely offered through a few monopoly organizations owned and/or operated by national governments. In today's highly competitive satellite industry there are hundreds of different companies offering one form or another of a service via many scores of satellite systems.
- Rapid Shifts from Co-Axial Cable to Satellites and now to Fiber Optic Systems. At the start of the age of satellite communications, space systems were many times more broadband and cost effective than international coaxial submarine cable systems. Developing countries largely relied to satellites to be connected to the world. Just as satellites replace co-axial cable for long haul interconnections, today's fiber optic cable systems tend to be much broader band and cost effective in being able to serve heavy trunk routes.

This does not mean that satellites are now an outmoded technology and somehow obsolete. Instead it means that the role of satellites has changed to complement fiber optic cable systems and thus to do what satellites do best. Thus the strength of satellites today is to provide broadcast services, to offer mobile satellite communications, and to allow connectivity to rural and remote areas. Satellites are also very good at providing highly distributed services for large enterprise networks, and special networking needs for defense-related applications when terrestrial infrastructure is simply not available. Terrestrial networks cover effectively only about 3% of the world's surface. Satellites are still important for the remainder of the globe.

All of these dramatic changes in the satellite world have meant that business planners and marketing experts have had to remain alert and highly responsive to changed market conditions. Some of the more important changes to the world of satellite business have included the following:

The Shift from "Wholesale" to "Retail"

In the early days of satellite communications the entities that launched and operated satellite networks saw their role as deploying space infrastructure and supplying the resulting capacity to governmental entities. They were clearly in the wholesale side of the business and not concerned with services and applications.

With the rise of competitive systems in the late 1980s and 1990s the shift from technology to markets became increasingly important. The rise of the Internet, VSAT, TVRO and handheld terminals and enterprise networks moving services to the "edge" all contributed to the shift. But it was particularly the opportunity to sell services directly to the end user with the rise of direct broadcast satellite networks and mobile satellite services that hastened the shift to retail sales and marketing. The opportunity for satellite service providers to offer services directly to the end user and for fixed satellite service providers to provide managed services to a growing number of corporate and government customers allowed the satellite industry to move up the "value added ladder" in the sales of their increasingly valuable services that contributed to the overall growth of the industry. During this shift the providers of direct broadcast satellite services and entertainment surged to the fore in terms of revenues. Today organizations such as DirecTV, Dish (Echostar), BSkyB, SES Global the Broadcast Service of Japan, etc., dominate revenues as the broadcast satellite services, based on direct retail sales, has far outstripped the fixed and mobile satellite service markets worldwide.[5]

The Shift from Technology to Applications

When satellite organizations were in the business of deploying space infrastructure they had a sharp focus on technology and developing the next generation of satellite capability. Intelsat and the Comsat Corporation had major research programs, and Comsat Labs led the shift to digital satellite technology. When the shift to competitive services and applications occurred, the satellite service providers shifted their focus to marketing and applications and left technology to the aerospace industry that manufactured spacecraft. The heads of satellite service organizations shifted from being technologists to being marketing experts and financial officers. Equity investment firms bought and sold satellite service organizations, and their focus was on new applications, market growth, profit margins, and key financial indicators such as debt to equity ratios. They understood that their customers were not interested in technology or infrastructure but new applications and services that could allow them to do their various jobs more quickly and efficiently.

The Shift Toward Globally Integrated Services

The initial days of satellite services in 1965 represented a heady period for the Intelsat organization. The new *Intelsat 1* with 240 voice circuits represented a facility with many times the capacity of the coaxial cables that crossed the Atlantic at that time, such as TAT-1 and TAT-2. Even with so-called time assignment speech interpolation (TASI) the largest submarine cable of the day had only 72 voice circuits.

By 1969 Intelsat had a global network of satellites each of which had 1,200 voice circuits plus two high quality color television channels. Submarine cables at the time could not even support high quality live color television broadcasts. The dominance of communications satellites, however, faded with the deployment of fiber optic cable systems beginning in the 1980s. Increasingly capable fiber links could provide not only a huge amount of throughput capacity but extremely high quality

---

[5] *Op cit*, Chartrand, pp. 263–64.

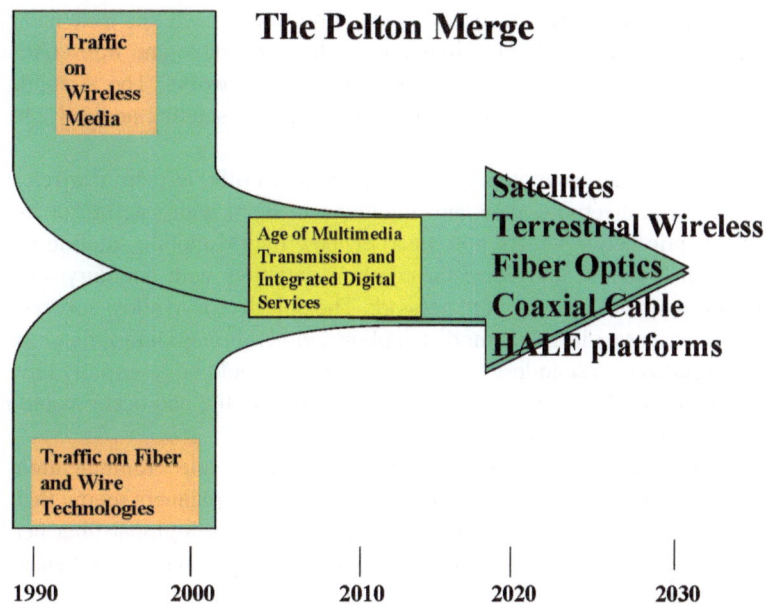

Fig. 7.1 The "Pelton Merge" and the trend toward digitally based multimedia transmission. (Graphic supplied by the author.)

and low noise links with virtually zero bit error rates and extremely low transmission latency as well. This made fiber an excellent mode for broadband Internet-related transmissions. Some Chief Information Officers were so impressed with fiber optic connections that they designed corporate networks that relied totally on these terrestrial connections. This was, however, a mistake, in that fiber networks can and do fail and satellites offer an excellent restoration capacity. Satellites still offer value in terms of being able to offer rapid restoration of service when "back hoe fade" (i.e., a dug up cable) or a submarine cable is torn loose by a fishing trawler. Satellites are also well suited to highly distributed corporate networks and services to rural and remote areas – as well as broadcast and mobile applications.

The truth is that most users are content for their communications or IT-based messaging to go into a cloud and then see or hear it come out of the cloud at the other end. In truth the job of a telecommunications engineer is to have compatible standards and interfaces so that coaxial cable, Digital Subscriber Line (DSL) over copper wire, fiber optic cable, Ethernet, broadband terrestrial wireless, satellite transmissions or high altitude platform relays can all interconnect across a global cloud and provide reliable service for any user. This move toward integrated systems was first identified as a major trend in the 1980s in an article in *Telecommunications Magazine*. (See Fig. 7.1.)[6]

---

[6]Joseph N. Pelton, "The Trouble With Predictions: The 'Negroponte Flip' 20 Years Later" On Magazine, Feb. 2009 www.emc.com/leadership/tech-view/trouble-with-predictions.htm

A number of telecommunications organizations recognized this truth and began to invest in hybrid networks that integrated submarine cable and transcontinental fiber optic networks with communications satellite networks. The strengths and weaknesses of fiber optic networks and various types of satellite networks are presented in Table 7.1 below.[7]

Others have begun to create integrated systems that offer terrestrial wireless networks in rural and remote areas and interconnect them with satellite or in some cases fiber optic networks. As noted earlier, new hybrid mobile satellite systems such as Light Squared and Terrestar combine satellites with ancillary terrestrial components to create an integrated network. The objective is to allow subscribers to conveniently access, via a dual mode telephone, either terrestrial mobile or satellite mobile as needed on a seamless basis. Urban service would be essentially terrestrial wireless and exo-urban and rural would be largely satellite and users would never need to think about which media they would be using.

The bottom line here is that the providers of global telecommunications networks today are beginning to think differently. Instead of planners seeing their mission as being that of creating a global satellite network or a global fiber network, these organizations are beginning to think in terms of creating comprehensive and "seamlessly interconnected" systems. In short today's planners are seeking to create reliable and flexible global telecommunications networks with diverse components and complementary technologies to create truly global coverage. Fiber, satellites and broadband terrestrial wireless are all seen as key components of such a global network.[8]

Increased Scale of Operation

As this shift toward applications occurred, the CEOs of various satellite service organizations began to seek ways for their organizations to not only offer the broadest range of services, but also achieve greater economies of scale and more globally integrated markets. This quest for economies of scale led to a number of mergers and acquisitions. These market consolidations were partly designed to expand into new markets and partly designed to achieve economies of scale. SES Astra, which had made large profits from their direct broadcast satellite business in Europe, began purchasing (or at least purchasing an interest in) overseas satellite organizations.

Today, SES Global has holdings on all continents and has ownership interest in some 50 satellites. Intelsat, after having been divested of some of its satellites as part of its privatization process through the formation of the New Skies Corporation in The Hague, Netherlands, managed to acquire its largest and most aggressive

[7] Joseph N. Pelton, "The New Satellite Industry: Revenue Opportunities and Strategies for Success", International Engineering Consortium, Chicago, Illinois.

[8] Joseph N. Pelton, *Future Trends in Satellite Communications: Markets and Services*, International Engineering Consortium, Chicago, Illinois, USA, pp. 97–108

Table 7.1 Strengths and weaknesses of satellite and fiber optic networks. (Graphic supplied by the author.)

Comparing Satellite and Fiber Characteristics

Capability	Fiber-Optic Cable Systems	Single GEO Satellite in a Global System	Single MEO Satellite in a Global System	Single LEO Satellite in a Constellation
Transmission Speed	10 Gbps – 3.2 Tbps	1 Gbps – 10 Gbps	0.5 Gbps – 5 Gbps	0.01 Gbps – 2 Gbps
Quality of Service	$10^{-11}10^{-12}$	$10^{-7}10^{-11}$	$10^{-7}10^{-11}$	$10^{-7}10^{-11}$
Transmission Latency	25 – 50 ms	250 ms	100 ms	25 ms
System Availability	93% – 99.5%	99.98% (C-Ku band) 99% (Ka band)	99.9% (C-Ku band) 99% (Ka band)	99.5% (C-Ku band) 99% (Ka band)
Broadcasting Capabilities	Low	High	Low	Low
Multicasting Capabilities	Low	High	High	Medium
Trunking Capabilities	Very High	High	Medium	Low
Mobile Services	None	Medium	High	High
Cost-Efficiency	Very High ($5 to $10 per year per transoceanic voice channel)	High ($20 to $50 per year per transoceanic voice channel)	Low (more than $100 per year per transoceanic voice channel)	Low (more than $150 per year per transoceanic voice channel)

Note: The rate of 3.2 Tbps assumes 40 monomode fibers with each being able to transmit 80 Gbps.
Source: Pelton, Joseph N., The New Satellite Industry: Revenue Opportunities and Strategies for Success (Chicago, Ill: International Engineering Consortium, 2002).

competitor, the PanAmSat Corporation. With this and other acquisitions Intelsat has also grown to control some 50 satellites around the world. Space Systems Loral has invested in Telesat Canada, XTAR and other ventures to diversify from being a satellite manufacturer to become a satellite service provider. The Boeing Corporation with its overseas venturing has done the same. Around the world partnerships, liaisons, mergers and acquisitions are now a routine part of the global satellite industry. Satellite manufacturers in pursuit of higher value service offerings and profitability have also sought global integration and increased scale of operation as well.[9]

Conclusions

The world of satellite communications is complicated enough just in terms of the wide range of technologies that are required to design, engineer, manufacture, test and launch a spacecraft and then seek to operate it in the hostile environment of outer space for periods that may be as long as 15 to 18 years. It is thus a considerable challenge to recognize that there are a host of other challenges and issues that must be mastered if one is to operate a satellite system successfully in today's world. These issues include the assignment of appropriate frequency spectrum by a national licensing agency, successful registration of a satellite system with the ITU International Frequency Registration Board (IFRB) and then going through the intersystem coordination procedures of the ITU. Subsequently one must also obtain "landing licenses" to operate communications satellite services in each and every country.

Despite new international trade regulations overseen by the WTO, it turns out in fact that obtaining such licenses and being able to compete on an equal basis with local satellite organizations can be quite a challenge. Sanctions to enforce open competitive measures under the GATS are often slow to be truly implemented.

One of the major challenges to the satellite industry is to cope with a great diversity of standards. It has been difficult for the satellite industry to optimize its operations so as to be fully compatible with Internet-related services, but most of these issues have been resolved with the creation of IPoS standards. The largest challenge of all has been the industry's adaptation to a host of major changes that have significantly restructured the industry and shifted its focus.

---

[9] Ibid.

The Future of Communications Satellites

8

Introduction

The satellite industry is exciting for a variety of reasons. It involves state of the art technology both in space and on the ground. It is international in scope. It is now a highly competitive industry with a rapidly expanding menu of services and applications. The market is highly dynamic, and staying on top of it requires a wide range of technical, operational, financial, regulatory, trade and standards-related knowledge.

The greatest challenge for satellite communications industry leaders is to be able to anticipate accurately future needs and then take action to deliver new capabilities, services or products as they are needed. Technical and operational trends are of particular importance since satellite specification, procurement, manufacture, deployment and operational implementation require significant lead times. Further allowance needs to be made for delays in the process as well as the possibility of a launch failure or satellite deployment problem. The development of new services and applications and anticipation of significant market trends, however, can be even more important since market demand needs to drive the procurement of new satellite facilities and not vice versa.

Key Satellite Technology and Market Trends

It is a mistake to believe that what is past is prologue to the future. A period of rapid growth can be followed by a period of recession and stagnation. A period of tepid demand may be quickly reversed by an appealing "killer app" or highly capable new technology. In the book by Steven Schnaars entitled *MegaMistakes*, the author presents numerous examples of how historical trend projections or elaborate methodologies to predict the future have often led to faulty forecasts.[1] Nevertheless some trend lines can form a dominant pattern for an industry for a consistent period of time – through times

---

[1] Steven P. Schnaars, *MegaMistakes: Forecasting and the Myth of Rapid Technological Change*, (1986) Simon & Schuster, New York.

J.N. Pelton, *Satellite Communications*, SpringerBriefs in Space Development, 95
DOI 10.1007/978-1-4614-1994-5_8, © Joseph N. Pelton 2012

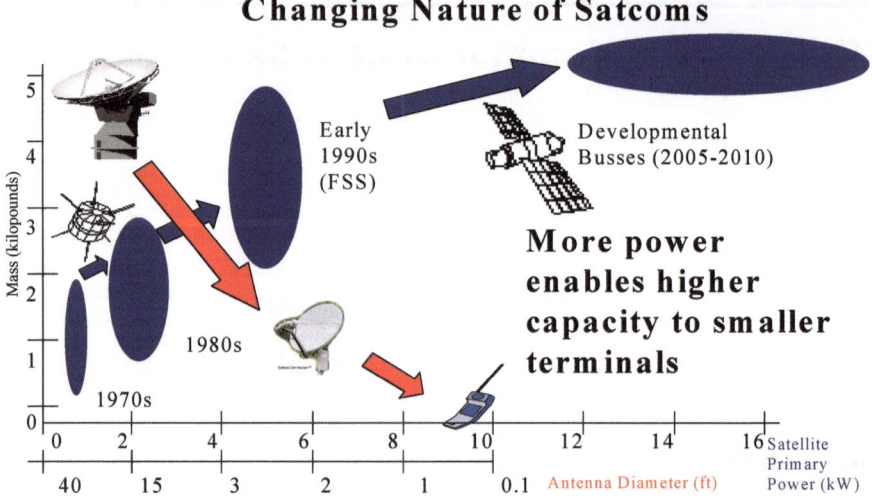

Fig. 8.1 Fifty years of technology inversion in the satellite industry. (Graphic supplied by the author.)

of boom and times of bust. Such a powerful "technology inversion" trend (as explained in the previous chapter) has dominated the satellite industry for a half century. As can be seen in Fig. 8.1 satellites have gone from being small to large and complex, while ground antennas have gone from large and complex to quite small. This trend is driven by the logic of finding a way to consistently decrease total systems costs in times of prosperity as well as in times of recession and consolidation.

If one examines the history of the Intelsat satellites and the Earth stations standards specifications, the trend toward higher capacity satellites and smaller ground antennas is dominant over a forty-year period. The growth of global demand combined with systems engineering consistently produced this megatrend that can be seen in the record that is manifested in the history of Intelsat satellites from 1965 to the present. But this is not an historical trend unique to Intelsat. This dominant pattern of development can likewise be seen in the systems with a long lineage such as Inmarsat, Eutelsat, Arabsat, etc. The technology inversion trend as shown in Fig. 8.1 thus seems to be a useful predictor of the future for some time to come.

There are no other similar dominant trends in the satellite industry that can be used to forecast the future. And even the technology inversion model seems in some ways to have run its course in terms of scale factors. One must begin to wonder as to how much larger satellite units can become and how much smaller can one shrink the size of user devices. There certainly are studies that have projected that "virtual antennas" (based on a swarm of pico-satellite arrays) could grow to square kilometers in size and other studies that have envisioned that there could be dual mode wireless and satellite units that might assume the form of an embedded chip within the human body.[2]

---

[2] Takashi Iida and Joseph N. Pelton, "The Next Thirty Years", in *Satellite Communications in the 21stCentury: Trends and Technologies*, edited by Takashi Iida, Joseph N. Pelton and Edward Ashford (2003) AIAA, Reston, Virginia pp. 169–199.

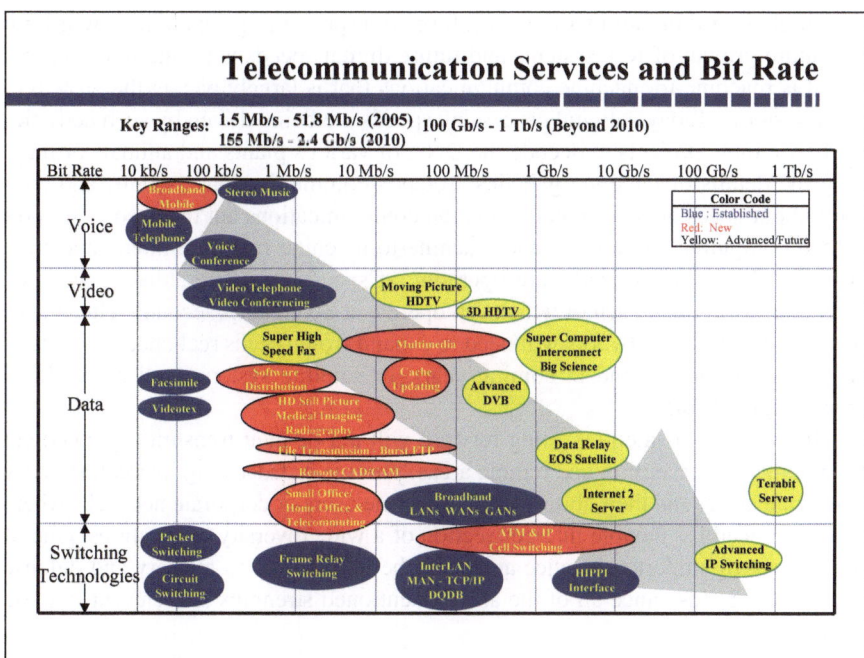

Fig. 8.2 The advancing digital speeds for ICT services in the 21st century. (Graphic supplied by the author.)

There are certain other trends that transcend satellite communications that are relevant to projecting the future of the industry. These trends relate to the overall growth and development of telecommunications and information technology (IT) and today often referred to as Information and Communications Technology (ICT). The advance of digital technology, the increased capability of digital processing and increasingly broadband systems with accelerating throughput capabilities is another historical trend line that has also been evident for many, many decades. The following graphic indicates just how dynamic the growth of digital services are and how rapidly ICT capabilities continue to expand. (See Fig. 8.2.) Blue in the graphic indicates established services, red indicates relatively new services and yellow indicates the most advanced and/or anticipated future services. The key element to note from this chart is not the specifics of any particular service that can vary from year to year but the relentless march of more and more services that are shifting from left to right – from slower bit rate services to ever-faster throughput applications.

The limiting factor here would seem to be the processing speeds of the human brain. Arthur C. Clarke once characterized the human animal as a biped carbon-based life form that processed information at about 64 kilobits per second. This is certainly about the rate that a typical human reads text, but when we watch television or a 3D movie the rate is probably a bit higher. The fact is that we process less than 1% of the information in modern movies or high definition television.

The flaw in thinking that human information processing capabilities will limit the future growth of ICT systems and future digital speeds is in failing to observe that it is machine-to-machine communications that is largely pacing the growth of these systems. Today's digital processing speeds are being driven by such activities as the Genome projects to decode the DNA of various plants and animals or mapping the details of far flung galaxies. Today's communications systems and their throughput needs are driven not by human communications but the need to support "cloud computing," Internet 2 and machine-to-machine real-time interconnection. Already fiber optic networks are operating at speeds of 3.2 terabits/second (i.e., forty strands of fiber, each of which can operate at 80 gigabits/second). This means that digital satellite networks will need to expand from speeds reckoned in gigabits/second to terabits/second in coming years, or satellite systems will be rendered increasingly obsolete.

The strength of satellites going forward will not be their transmitting speed but their ability to serve rural and remote areas, provide broadcasting interconnectivity over wide areas, and their ability to serve mobile users or corporate networks where there is a need for flexible interconnection of a wide diversity of localities. One of the continuing important service areas will be with regard to military and defense related ICT needs, since all of the above mentioned strengths are relevant to these operations.

Market Trends Related to Military and Defense-Related Satellite Services

There have been military applications of satellites in operation since 1965. Indeed it was military projects that deployed some of the world's first experimental communications satellites such as SCORE (1958) and Courier 1B (1960). The initial defense satellite communications system began operations the same year the first commercial satellite system, Intelsat, was deployed in the mid 1960s.

The logic of having a separate military communications satellite system has several key components. Commercial systems are designed to serve heavily populated areas, while military conflicts are often in rural and remote areas or involve littoral areas off the coast of countries or the high seas. In short, military satellite systems are designed for geographic coverage of areas not seen as prime real estate for commercial systems. Secondly, defense-related communications systems are considered as potential targets for attack, and thus military satellites tend to be designed with radiation hardening, anti-jamming and other special protective features. Thirdly, military systems are also designed with special encryption capabilities to protect against secret messages being intercepted by hostile forces. Finally, there is the question of having desired communications capacity on demand when hostilities or a special communications need such as disaster relief suddenly arises. Since commercial satellite systems are as heavily loaded as possible to achieve the highest level of revenues, there is no particular reason to think that spare capacity to respond to a disaster or a conflict will be available on demand.

For these reasons there are a number of defense-related satellite systems that are in orbit to serve military needs. There are particular systems available to meet the needs of the United States and NATO nations, but there is a growing capability by other countries such as Russia, China, and Japan that are deploying dedicated military satellite systems. Russia during the time it was the Soviet Union had military communications satellite capabilities at a relatively early period, but these facilities were unique in that they provIDed military, government and other telecommunications services in the "public sector" on an integrated basis.

The next chart shows that reliance on commercial satellite systems remains a part of the plan to meet overall military communications needs. Some requirements such as providing television sports and entertainment to overseas military personnel do not require specialized facilities. Certainly commercial systems that lack special protective requirements can provide more cost effective service. In the case of the U. S. defense communications planning, up to 20% of requirements are conceived of as being potentially met by so-called "dual use" of commercial systems. Today commercial systems such as Intelsat, Inmarsat, Iridium, Globalstar, Eutelsat, as well as several more are extensively used for such dual use applications.[3]

In places of current conflict such as Afghanistan and Iraq and other parts of the Middle East commercial systems are extensively used to support military-related communications. For the most part commercial satellite systems are used for entertainment, sports and news broadcasting and for communications with troops for a variety of purposes. Nevertheless, dual use facilities are not employed for tactical purposes in direct support of military operations or for purposes of maintaining secure and highly encrypted communications and combat operations.

There are international treaties that call for the non-military uses of outer space and these treaties often ban the operation of military weapons in outer space or other celestial bodies. In particular, Article IV of the Outer Space Treaty explicitly bans the use or deployment of nuclear weapons in outer space. However, this treaty is silent on such issues as the deployment of military communications satellites, laser systems or anti-satellite technology. These issues are currently under discussion in Geneva, Switzerland, through such agencies as the U. N. Institute for Disarmament Research (UNIDR).[4]

The demands of defense-related communications are not as clear-cut as the above chart (Fig. 8.3) would suggest. This is because it is never clear what mix of broadcast, fixed, mobile and data store and relay services may be needed at any one point in time. Just one example of this uncertainty is the demand related to the use of UAV drone aircraft to undertake surveillance in military or disaster relief areas. The video that is generated from such optical reconnaissance by drone UAVs has served to drive upward greatly the use of commercial communications satellites in the past

---

[3] Joseph N. Pelton "Satellite Performance and Security in an Era of Dual Use" On-Line Journal of Space Communications, spacejournal.ohio.edu/issue6/pdf/pelton.pdf

[4] Ram Jakhu, John Logsdon, and Joseph Pelton, "Space Policy, Law and Security", edited by Joseph N. Pelton and Angelia Bukley, *The Farthest Shore: A 21stCentury Guide to Space*, (2009) Apogee Press, Burlington, Ontario, Canada, pp. 291–320.

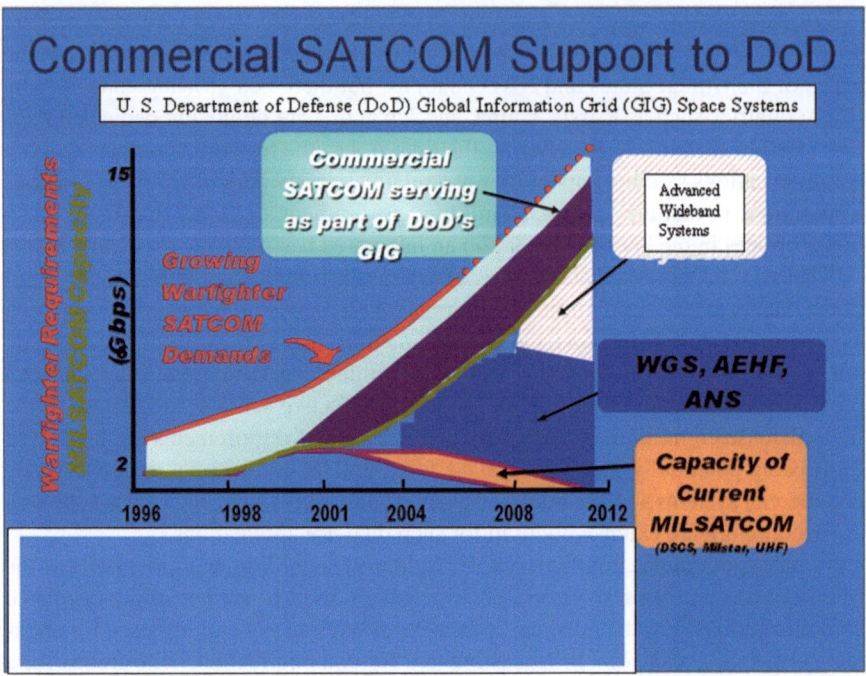

Fig. 8.3 Projected use of commercial "dual use" satellite system by U. S. Department of Defense. (Graphic supplied by author.)

few years. Mobile satellite use has also expanded enormously in volume as well. It is for these reasons that the projected stable requirement of about 5 gigabits of commercial satellite for the dual use capacity in future years may well be an underestimate of total usage volume.

New Trends in Mobile Satellite Service

The field of mobile satellite services has evolved significantly in the past quarter century. Services started with relatively low powered satellites for maritime and aeronautical services. MSS satellites are today much different from the first generation Marecs and Marisat satellites of the 1980s. Current MSS networks, just like BSS networks, have become ever more powerful.

The deployment of more powerful satellites with larger aperture antennas was first undertaken to serve maritime and aeronautical markets more effectively. The key surge of development, however, came in the late 1990s in response to the desire to provide land mobile services and to compete with cellular mobile services – at least in rural and remote areas. The first of these attempts to provide global mobile services came via low Earth orbit constellations such as Iridium and Globalstar. These LEO-based satellite systems have now been joined by extremely wide aperture satellites operating from GEO. Today second generation Iridium and Globalstar systems in LEO are competing with GEO satellite systems with very large aperture MSS systems that have antennas that are some 12 to 22 m in diameter.

What is quite new about the very latest mobile satellite systems is the idea of creating "hybrid" systems. The idea for these systems is to combine, in a seamless manner, terrestrial cellular systems with MSS space networks. The object is to provide mobile connectivity in rural, exo-urban and suburban locals via space systems with terrestrial cellular inside the city. This new service is called in the U. S. Mobile Satellite Service with Ancillary Terrestrial Component (MSS-ATC). The 18-m Terrestar and the 22-m Light Squared satellite networks represented this very latest type of mobile satellite configuration. This type of service, called MSS with complementary ground component (MSS-CGC) in Europe, will likely proceed forward – or not – based on the success of such systems in the United States. Another key issue in the US is the matter of frequency interference between the Light Squared system and the immediately adjacent GPS frequencies used for space navigation.

The real question for the future is just how large a deployable space antenna can be designed and built economically and reliably. The 22-m Light Squared MSS-ATC satellite built by the Boeing Corporation (with the unfurlable antenna built by the Harris Corporation) has experienced difficulties in the deployment of its huge antenna. After "shaking" the satellite platform by use of onboard thrusters this huge antenna was fully deployed in December 2010. There have been future-oriented studies, however, that have suggested that 100-m antennas (using phased array feed technology) might be possible and that an array of pico satellite units could grow to square kilometers in size.[5]

Changing Market Demands and Sources of Revenue for Commercial Satellite Systems

The world of satellite communications has been in a state of flux for at least a decade, and the next decade does not appear likely to be greatly different. There has been a swirl of change in the ownership of satellite firms by equity investors and even larger changes in satellite markets due to the diversity of digital offerings related to video, voice and data services. Key to stirring up and sometimes muddling satellite markets has been a spate of new services that includes digital high definition television (including 3D television), digital video broadcasting applications for data distribution as well as voiceover IP and multimedia over IP. These digital services have redefined many satellite markets. The regulatory lines that divided direct-to-home television from broadcast satellite services or direct broadcast satellite from fixed satellite services have never been straightforward.

Today these lines of division seem even more fluid than ever before. Indeed the future seems to be moving toward large multi-beam satellites that can provide a

---

[5] *Op cit.* Iida and Pelton pp. 180–199.

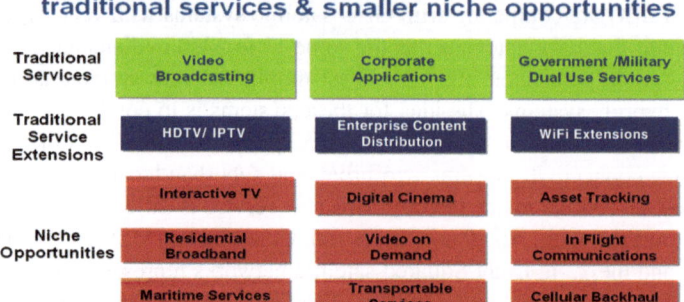

Fig. 8.4 Existing and new market expansions for satellite communications services. (Graphic by the author.)

multiplicity of services from multi-use platforms. The ITU process of allocating different frequencies for different services can often lead to inefficient use of spectrum and can be economically inefficient as well. Today there are fixed satellite systems that are used to provide video services to ships at sea, and broadband mobile satellite services are providing fixed communications to broadcasters. Until multipurpose frequency allocations for satellite communications become possible the crossover use of different bands for different services will continue. What is clear is that there is a wide range of service growth opportunities in all areas of the FSS and BSS markets, starting with so-called traditional services and expanding into new service offerings – some of which are noted in Fig. 8.4 below.[6]

There is another whole dimension to satellite markets that goes beyond these new and expanding service opportunities for satellite communications. These new opportunities are really not new markets or services, but rather they are a new market opportunity to provide complete "turnkey" end-to-end services to large customers. These customers may be military or governmental agencies, or large multinational enterprises or broadcasters or other substantial users of satellite services.

The largest single market expansion today is in the area of so-called "managed services." This is where an entity contracts to provide the full range of end-to-end services on a 24/7 basis. A single entity undertakes to provide the complete network of satellites, ground systems and monitoring (i.e., troubleshooting, repair and maintenance). This entity can thus design and implement a satellite system on a fully integrated basis and likely benefit from economies of scope and scale in the process. Such restructuring of markets allows providers of mobile and fixed satellite services to move away from "wholesale" provision of space infrastructure and migrate toward "retail" offering of a full menu of services to end users. Intelsat General now does this, and Inmarsat is being restructured to do so. Many smaller organizations

---

[6] Joseph N. Pelton, *Future Trends in Satellite Communications Markets and Services*, (2005) International Engineering Consortium, Chicago, Illinois, Chapters 5,6 and 7.

such as Cap Rock, Tachyon, Segovia, CommSystems, etc., have been providing such services for the U. S. military for a number of years, but today this is trending toward a major market shift to support a much wider range of satellite customers.[7]

Integration of Existing and New Global Telecommunications and IT Networks

The future of satellite communications will remain bright as long as satellites remain uniquely qualified and highly cost effective in terms of providing broadcasting and multicasting applications, mobile applications, dual-use applications for defense-related services and rural and remote applications. Digital video broadcasting services that support the provision of digital information to the "edge" in support of distributive networks for large organizations will remain a key strategic service for satellites for some time to come.

Despite these strategic strengths, it will remain the case that satellites will continue to be adjuncts to large terrestrial ICT networks based on a fiber optic network backbone. Satellites, broadband terrestrial networks and fiber optic networks will increasingly be integrated into unified networks – often with common ownership or closely interlocked partners. To the extent there is consolidation of satellite system ownership and extension of "managed network" operations this will likely serve to accelerate the consolidation process and to facilitate the planning of large-scale information networks.

Today these ICT networks and their ownership are concentrated in Europe, North America and Japan. Tomorrow this balance will shift as the global economy shifts so that so-called developing and industrializing countries will have a larger ownership share in global ICT networks and a greater say in the design and configuration of both terrestrial and satellite systems. The global economy has shifted sharply in recent years and projected to shift even more sharply still.

Up until about five years ago the global economy was about 60% controlled by the United States, Canada, Japan and Europe, but this is projected to shift by 2050 to about 25% of the global economy being directed by these countries. Meanwhile the rest of the world's economies will shift from about 40% of the total to as high as a 75% share. As this economic shift occurs the ICT networks will likewise shift in these directions as well and potentially at an even faster rate. Because of the emphasis on mobility in modern economies a significant portion of these new networks will likely be based on terrestrial broadband wireless and satellite technologies, albeit with a fiber backbone.[8]

---

[7] Intelsat general satellite solutions, http://www.intelsatgeneral.com Inmarsat managed network services http://www.inmarsat.com

[8] James Wolfensohn, (former President of the World Bank), *Washington Post*, November 14, 2008, p. A 19.

Conclusions

The future of satellite communications remains based on the extension of digital systems both within existing markets and new applications based largely on IP technologies. Managed networks and large-scale integration of ICT markets will be a major part of the future. Dual-use applications in support of defense-related applications and the drive to have access to broadband mobility will also drive future market trends. The motive force of satellite "technology inversion" has yet to run its course. The need to deploy very large-scale and powerful multi-beam satellites will assist the movement toward integration of international ICT networks. Satellites will thus remain important, but they will continue to be the ancillary part of large-scale fiber and broadband terrestrial networks that will dominate new communications and IT investment in the decades ahead.

Top Ten Things to Know about Satellite Communications

9

Introduction

There are, of course, a great many things to know about the field of satellite communications. These range from the technical, operational, financial, and business aspects to the regulatory environment. The satellite communications markets and activities are quite dynamic and vary widely around the world and thus require constant monitoring. Also risk management is a key element of maintaining a successful operation in the field. The following top ten things to know about satellite communications are drawn from the preceding chapters and interviews with many people in the field from around the world.

Technology

#1. Satellite communications technology, both in terms of satellite systems and ground systems, continues to evolve rapidly around the world. Continuing research and development, however, will be needed in order for satellite markets to continue to expand, applications to increase, and satellite costs to decrease.

Rapid strides to improve the transmission technologies for fiber optic and broadband terrestrial transmission systems have placed additional pressure on satellite communications technology to advance rapidly and to be as competitive as possible. Current satellite technology development goals include: (a) improved multi-beam antenna systems to support the creation of higher-powered spot beams. This is coupled with the development of more capable onboard processing and switching systems that allow for more effective interconnection of these beams; (b) improved satellite communications subsystems optimized to operate reliably at higher frequencies, especially the 30/20 GHz Ka bands and even higher Q.V (48/38 GHz) and W (60 GHz) bands. This R&D is focused on new processing technologies and high-powered spot beams to overcome rain attenuation problems at these higher frequencies. This research includes developing new abilities to operate reliably (especially in areas experiencing heavy rainfall and possible loss of signal). Research areas

J.N. Pelton, *Satellite Communications*, SpringerBriefs in Space Development, DOI 10.1007/978-1-4614-1994-5_9, © Joseph N. Pelton 2012

include finding ways to supply additional power margin in affected beams, longer dwell times in CDMA and TDMA systems and slower transmission throughput in order to overcome increased bit error rate; (c) improved IP over satellite standards and techniques so that satellite transmissions involving broadband Internet and TCP/IP connections can work more efficiently and overcome the effects associated with satellite transmissions delays; (d) improved satellite antenna technology (i.e., phased array antennas and feed systems and highly conformal and large aperture deployable parabolic antennas); and (e) constantly improved user antennas and transceivers, including satellite handheld phones with improved ASIC (application specific integrated circuits) technology to allow the design and manufacture of lower cost and smaller and more versatile handsets and micro-terminals (VSATs and USATs).

These are, of course, only some of the more important research objectives for satellite communications. Other key technical areas are noted in Chapter 2. Technological R&D is also being pursued to achieve more cost effective ways to launch satellites into orbit, to develop improved inter-satellite links, to reduce satellite in-orbit operating costs via autonomous operation, and to generally make satellites more cost efficient, reliable, and better integrated with terrestrial telecommunications systems.

Finally, in addition to developing new technology and seeking enhanced reliability and throughput capability, there is an ongoing effort to find ways to manufacture satellites and Earth stations more effectively and with lower costs and testing without sacrificing reliability and mean time to failure. At this time many civilian space agencies worldwide, including France, Italy, Germany, Japan, Canada, Brazil, India and China, remained focused on R&D to support advancements in satellite communications. In the United States, NASA has largely stopped its research program in satellite communications, but military R&D units maintain an active research program in this area.

#2. Sustained frequency allocations and interference mitigation from other satellite systems as well as from terrestrial communications systems, and in the future high altitude platform systems (HAPS), will likewise be critical for the success of satellite communications.

The future of satellite communications is heavily dependent on continued allocation of wide band spectrum to support the future growth of communications satellite services. In recent years there have been few new allocations to satellite communications except in quite difficult to use frequencies in the extremely high frequencies (EHF) above 30GHz. Further there have been increased instances where satellites have had to experience sharing of their allocated frequencies with other types of use – either terrestrially or with regard to the new high altitude platform systems (HAPS). These are communications platforms that fly in the stratosphere. The entity responsible for global coordination of frequency allocations is the ITU in Geneva, Switzerland, a specialized agency of the United Nations that addresses all forms of telecommunications and broadcasting including the formal allocation of use of the radio frequencies. In the military and defense-related bands there has been some erosion of allocations for satellite communications and radar-related service.

The other major issue that the satellite communications field faces with regard to the use of radio frequencies is the increasing problem of interference. At the outset, communications satellites in geosynchronous orbit were spaced further apart to minimize inter-satellite interference. As the demand for service rose satellites in GEO (or Clarke Orbit) have moved closer and closer together. In addition the aperture size of user antennas has decreased to allow the cost of ground systems to decrease, to make antennas more compact and mobile, etc.

The decreased aperture size of ground systems creates more harmful interference from side lobes. There are now organizations such as the Satellite Users Interference Reduction Group (SUIRG) as well as various working groups formed within the ITU to develop standards and operating procedures to reduce harmful interference. As new satellite systems are filed with the ITU in terms of frequency and orbital location, these filings are internationally shared among all ITU members to allow an assessment of interference and to seek ways that such interference might be mitigated by improved use of digital signaling and encoding, different transmission plans, or alterations in the satellite design.

Today there are some 800 active satellites in orbit, and virtually all of these involve some form of data transmission and information relay. Well over 300 communications satellites for civil and military purposes operate primarily in GEO, MEO or LEO. Although harmful interference has been minimized in a number of ways, the problem today is a serious one. The sharing of frequencies by communications satellites with terrestrial networks, HAPS, and, of course, other satellites, remains a major challenge going forward. Improved digital transmission and encoding technology and improved satellite antenna technology has served to mitigate these interference problems in major ways.

#3. Launch services remain a key barrier to lower satellite communications costs. Breakthrough technologies that would allow lower cost access to outer space and that might allow low cost refueling, retrofit or repair of communications could aid satellite communications cost efficiencies enormously.

Over the past several decades communications satellites have become hundreds of times more powerful. Frequency reuse has expanded greatly effective access to more spectra. Satellite lifetimes have been greatly extended – often to last 10 to 15 years or more. Space antenna systems have become hundreds of times more efficient in terms of their ability to create powerful spot beams and to efficiently interconnect these beams.

The result has been to develop satellites that are hugely improved in terms of power, capacity, lifetime, performance and cost efficiency as measured in channels per kilogram in orbit. On the launch services side, rocket launchers are more reliable. Further, large launch vehicles when providing multiple launching of satellites at one time are more cost efficient. But in truth, when corrected for inflation, launch rates have at best shown marginal improvement in cost efficiency. It is hoped that new and improved launch technology, or some entirely new lift capability such as using tether technology, space elevators, rail guns or some now unknown concept can evolve to allow the launch of satellites to become significantly more reliable and cost effective. Those who look well into the future have even envisioned that one day satellites might be assembled using processed materials and automated machines on the lunar surface. In such a future, satellites might be in effect "lowered" from the Moon to Earth orbit.

Markets

#4. There is really no single integrated satellite market. The major submarkets are video broadcasting (i.e., direct broadcast satellite services), fixed satellite services (including video distribution), mobile satellite services and data distribution (or machine to machine [M2M] services. Perhaps the important trend, however, is for large satellite service providers to offer not only chunks of satellite capacity (such as transponder leases) but also to offer "managed end-to-end services" to an ever widening range of users.

There are many books written about the field of satellite communications and all such space systems, largely using similar technology in terms of power, stabilization and pointing capabilities, antennas, tracking, telemetry and control, and launcher services. Certainly some go into different orbits and have different shapes and sizes, but at first they may seem quite similar. In truth, the markets are today quite differentiated from one another. In short, there is no such thing as an integrated satellite services market. The direct broadcast satellite service market is by far the largest, is closely tied to the entertainment industry and involves the direct delivery of a service to consumers. It thus follows its own market trends and tends to have separate regulatory controls and to be much more of a national or regional type of service.

Mobile satellite services are largely tied to maritime and aeronautical services at the global level, but when it comes to broadband cellular type "land services" for consumers, the industry today is increasingly tied to national and domestic markets. Land-based cellular and mobile satellite services are tending to be merged into integrated companies, as represented by Terrestar and Light Squared in North America and other similar-type projects in Europe and Asia. Although mobile satellite services are the smallest of the big three (broadcast, fixed and mobile) in terms of revenues this is currently the satellite service sector that is growing most rapidly in terms of percentage growth. This is, in part, because mobile satellite services are making the transition from being a wholesale business, selling to telecommunications companies, to becoming a retail business that markets and sells directly to consumers.

Fixed Satellite Service (FSS) was at one time the only game in town, but today this is a much smaller market than broadcast satellite services and it remains the one market that largely sells its services to telecommunications companies, broadcasters, cable television systems or large corporations. Its inability to make the transition from a wholesale to a retail enterprise, with a consumer clientele, has limited its growth and market expansion. It also functions as more of an "extension" of telecommunications, broadcasting, and cable television companies plus military and governmental specialized networks, in terms of its business planning. In vernacular terms, FSS service operators tend to be the "tail" and not the "dog" in defining its own strategic mission and controlling future patterns of market growth.

Although digitally based Internet services (i.e., TCP/IP-based virtual private networks) offer the opportunity for FSS operators to connect directly to small offices and home offices (i.e., the SoHo market), its near-term prospects remain largely wholesale in character. For the longer term prospects will be brightest if FSS operators can also migrate more toward retail sales to individual users and business clients, as has been the case with direct broadcast satellite and mobile satellite services. Turnkey service is one current strategy for the FSS industry to move to a more retail type of operation with an ongoing revenue stream from end users.

Finally there is the much smaller data distribution market, or (M2M) satellite services. This type of service is often aligned with space navigation services since the main clientele are those wishing to keep track of large fleets as well as to send and receive short data messages to mobile vehicles, ships, aircraft or people. Organizations such as car rental agencies, shipping lines, railway companies, trucking lines, etc., are the users of this type of service, plus people traveling to rural and remote locations. The character of this market does to seem to afford major future growth opportunities.

Today these four satellite communications markets follow different growth patterns and have different technological needs to support their future growth. In the future multipurpose large-scale satellites designed to provide different services in different frequency bands might allow re-integration of some of these diverse satellite communications markets, but this is not likely to happen in the near-term and perhaps never will occur because of the diversity consumer and market needs.

Perhaps the most important trend for mobile and fixed satellite service providers is the increased ability to offer "managed end-to-end" services to an ever-widening pool of end users. For some time the direct broadcast satellite providers have offered retail services to end users. For this reason they have been able to ascend higher up the value-added ladder and achieve higher profitability. This lesson has not been lost on those in the mobile and fixed satellite service industries. Increasingly large satellite service providers have found ways to deploy complete end-to-end managed networks to end users and become much more of a full service provider within a retail market. Today 20% to 30% of revenues for large satellite service providers in the mobile and fixed services markets come from full service offerings, and these percentages of total revenues continue to grow.

#5. A major element of the global satellite communications market is the so-called "dual use" services. These are the provision of military or defense-related telecommunications services by commercial satellite providers.

There are numerous military and defense-related satellite communications systems operating in the UHF, X-Band plus very broadband systems operating in the EHF bands. These systems are primarily operated by the United States and European countries, although other dedicated systems are now emerging. Despite the large capacities and diversity of these systems there has been an increasing use of civilian satellite systems to augment the capacity of these military and defense-related systems. Many of these applications are quite similar to conventional civilian usage such as distribution of television programming to overseas troops or provision of capacity for military personnel to send e-mail to family and friends.

Civilian satellite systems, because they do not require special features such as radiation hardening, high level encryption, etc., are lower in cost and can provide these types of services at lower cost. Although civilian satellite systems are not used for "tactical communications," such as the targeting of weapon systems, they are also used to augment communications in war zone areas. Currently capacity on the order of 5 gigabits/second is derived from commercial satellite systems, and if anything those requirements are growing. In a parallel trend some countries, particularly in Europe, have developed specifications for defense-related type requirements and then through competitive procurement processes arranged to lease capacity to meet these needs. Under these arrangements the commercial partner is allowed to lease additional spare capacity to commercial users. This type of arrangement, that was first used by the U.S. Navy under its Leasesat program in partnership with the Comsat Corporation, is now increasingly in use in Europe.

#6. Digital communications satellite services are critical to satellite communications markets and their ability to continue to expand. Coding and other more efficient use of available spectrum are more important to satellite communications (in order to be able to transmit more bits per hertz) than is the case for fiber optic services because satellites are severely "frequency limited" while fiber optic networks are not.

Today's communications satellites are in most cases specialized digital processing systems in the sky. The exact function of broadcast, fixed, mobile or data relay satellite systems is determined by the specialized software programming that is loaded onto the satellite processors. In essence satellites are software-defined digital equipment despite the very special hardware that must be included on a satellite to allow it to survive and operate for extended periods of time in outer space.

Although there have been many advances in satellite technology, such as improved power and antenna systems, the greatest improvements in satellite performance in terms of transmitting more bits of information per available bandwidth (i.e., bits/Hz) has come from more efficient software and improved encoding systems such as "turbo coding." Twenty years ago a 72 MHz satellite transponder could transmit perhaps two analog color television channels of medium quality. Today that same 72 MHz transponder might support 14 or more digital television channels through the use of the most advanced MPEG digital standards for compressed television channel signaling. Continuing advances involve being able to compress information through more and more efficient encoding. In short, the most promising pathway to increased satellite performance and cost efficiency is through enhanced software and improved digital encoding.

There is, however, a problem for satellite communications in terms of very intensive use of coding. In the field of satellite communications one cannot simply use more and more efficient encoding to send more and more information. Why not? The catch is that the noise, or interference, as measured in terms of bit error rate (BER), limits the efficiency of encoding; this is a particular problem for satellites. In a fiber optic cable there is virtually no interference and terms like "zero BER" or near- zero BERs of 10-12 are commonly used to describe a fiber-based transmission. In the case of satellites such performance might be achieved in clear sky conditions, but in higher frequencies and during rainstorms performance decreases and BER increases considerably. These conditions allow only limited encoding.

Ironically, fiber optic systems using high efficient Dense Wave Division Multiplexing have access to almost unlimited spectrum and thus specialized codec (coder/decoder) operations to send more bits/Hz of information; but they do not really need this, and it constitutes a needless additional cost. To keep pace with fiber optic networks satellite services need to develop highly efficient encoding systems because of limited frequency allocations. The special incentive is to find ways to do this without having signals interrupted or deteriorated by rain attenuation or other factors that degrade performance and drive up BERs in the transmission channel. Onboard processing, as noted earlier, is one key to achieve this basic objective.

#7. Satellite communications are strategically best positioned for the longer term to provide mobile satellite services, connectivity services to rural and remote areas, and broadcasting and wide area networking services. This is because these are areas where satellites can be superior to fiber optic networks from a technical, operational or financial perspective. Satellites can shine when very broad areas are to be served, traffic volumes are small, or where it is not possible or convenient to have fixed connections (as in the case of mobile services or where service is over deserts, jungles, mountains, or oceans).

In many ways, however, satellite and terrestrial networks can best be seen as complementary rather than necessarily competitive systems. This is because satellites are often strongest where fiber optic networks are weakest and vice versa. In the age of the "Pelton Merge" the best, most cost effective, and most reliable networks are likely to be a blend of satellite and terrestrial technologies to provide services to consumers that are "seamlessly" connected to end users based on open standards that allow space and terrestrial services to switch from one media to another on a virtually invisible basis.

There are a number of systems aimed at using a combination of satellite and terrestrial wireless services to provide Internet-based broadband services to rural areas and developing countries. The so-called O3b (for Other Three Billion) satellite system, a constellation of MEO satellites, is but one example of this type of merged terrestrial and satellite network. The U. S.-based hybrid satellite and cellular systems, i.e., the Terrestar and Light Squared networks, represent such a combined terrestrial and space solution for 4G-type services in developed economies. A number of telecommunications carriers in the United States rely primarily on fiber and coax terrestrial networks, but also market their services in rural areas via satellite direct access to VSAT micro-terminals. Revenues for this type of service in the United States recently rose sharply to top $1 billion (in U. S. dollars) in 2009.[1]

---

[1] *Op Cit, Satellite Industry Association, 2010 State of the Satellite Industry Report.*

Business and Finance

#8. Economies of scale are extremely important in the satellite industry. This applies to the launch services and satellite and Earth station manufacturing side of the business and perhaps even more so to the satellite communications services businesses.

It is ironic that it was the desire for competitive economies of scale within the satellite services industry that led to the breakup of global or regional "monopoly organizations." These were entities operated as intergovernmental agencies empowered and essentially funded by governments, such as Intelsat, Inmarsat and Eutelsat. It has been the economics of scale and global marketing that have subsequently allowed a few "super satellite carriers," mostly owned by equity investors, to dominate global satellite services.

At the domestic or the regional level, in the case of direct-to-the-home satellite broadcasting services, only a few competitors have been able to sustain themselves either due to competitive pressures or acquisition by competitors. Economies of scale are achieved not only through larger and more cost efficient satellites and heavier lift launch vehicles but also through worldwide marketing systems and automation of operations and manufacturing capabilities. For the year 2009, for instance, total revenues for satellite communications rose to over $160 billion (in U.S. dollars) (i.e., services, satellite manufacturing, launch services and Earth station sales). This represented an 11% net increase in total revenues over 2008, but total employment in all of these various sectors decreased and represented an average shrinkage of jobs of 5.5% across the board.[2]

#9. Risk management has played and continues to play a key role in the satellite industry. This is because of the many risks associated with possible catastrophic losses and the rapidly changing and evolving global telecommunications markets.

There is the obvious case of the initial satellite launch and or the possibility of loss of satellite functionality in the harsh environment of outer space. Since the satellites are in orbit, often in GEO, which is one tenth of the way to the Moon, there is limited ability to repair enormously expensive failed spacecraft. Finally the satellite industry competes in a very difficult marketplace where spacecraft and launch arrangements must be completed years in advance while the marketplace and the offerings of competitors – in space and on the ground – are constantly shifting on a day to day basis and mergers and acquisitions have become commonplace.

There have been significant swings in the costs and form of risk mitigation available within the global satellite markets over time. In the very early years "launch insurance" was not possible to obtain at all. Then some insurance risk coverage began to be available, but it hinged on achieving coverage only after two consecutive launch failures, and even then coverage was at a relatively high cost. As the industry matured and actual claims against launch failures turned out not to be too severe,

---

[2] *Ibid.*

some of the largest organizations were able to obtain coverage at progressively better rates. The insuring firms entered the reinsurance market and thus no single insurance firm ended up with an excessively large exposure to a single failure.

A large satellite operator, such as Intelsat, with the largest number of launches with launch insurance coverage and good success rate, was at one time able to obtain launch coverage in the mid 1980s for only about 8% of the amount of the insured risk for a particular launch. Then a series of large claims occurred in a short period of time. In addition there was also a spurt of new entrants into the field by a number of different types of satellite operators for domestic, regional and global services. The net result was a sharp escalation of the cost of launch insurance coverage that surged upward to the range of 15% to 25% of insured risk. New entrants with only one or two satellites to be launched for a domestic system and with no previous engineering experience of monitoring the reliability of launch arrangements ended up paying the highest rates.

Likewise the cost of insuring in-orbit risks to satellites (including coverage against in-orbit collisions) also began to rise. Liability coverage against a major launch failure in urban areas are high even though many governments such as the United States have enacted legislation to provide additional levels of insurance protection against a catastrophic launch failure such as a launch vehicle from Cape Canaveral flying into Miami, Florida.

The bottom line is that risk management and risk mitigation remains a difficult and challenging part of the satellite communications industry. Changing safety standards, regulatory shifts and the statistical changes in actual performance in terms of launch or satellite failures make this an always changing landscape. The most recent factor that has begun to impact in-orbit and launch insurance is the steady increase in orbital debris and the actual recent collision that occurred between an Iridium LEO satellite and a defunct and out of control Russian Kosmos satellite

---

Trade and Regulatory Concerns

#10. The satellite industry is subject to many different types of international technical standards as well as numerous local, regional and international regulations, related to both trade and telecommunications services. It is also limited by available frequency allocations for various types of services. None of the regulatory or trade conditions are fixed and indeed are constantly evolving and changing. In addition there is also new concern about orbital debris and the "sustainability" of space due to increasing "space junk" that has accumulated in Earth orbit.

In the world of rocket science the only thing that is perhaps more challenging than the technology is the complex scope and changing nature of global regulatory oversight, frequency allocations, safety standards and trade guidelines and restrictions in the field of satellite communications. Entities that operate global satellite systems must deal with "landing rights" in perhaps as many as 200 different countries and territories around the world. Agencies that operate global or regional satellite services must actively monitor and in a number of cases participate in the activities of a large number of organizations that include but is not limited to: the ITU, the International Electro-Technical Commission (IEC), the International Standards Organization (ISO), the IEEE, the World Intellectual Property Organization, The WTO, the International Maritime Organization, the International Civil Aviation Organization, the U. N. Committee on the Peaceful Uses of Outer Space as well as the major regional or national technical standards organizations such as ANSI of the United States (as well as U. S. military technical standards), the European Technical Standards Institute, etc.

Although there has been some attempt within Europe and North America to streamline the filing of satellite systems and seeking approval of active Earth station arrangements in terms of frequency oversight, these trends have largely been offset by other regulatory or trade problems. Today there are still problems presented by "paper satellites," limited new frequency spectrum available for future satellite services, an increasing amount of orbital debris, and only limited success with regard to creating truly free trade for competitive satellite communications services within the WTO structure. In addition to these trade and regulatory issues, one must also recognize that the ever growing number of application satellites in operation, just by their sheer number, adds to the overall complexity and difficulty of space systems operations.

Appendix 1: Key Terms and Acronyms

AM Amplitude Modulation.

ANSI American National Standards Institute.

antenna A device that receives or transmits a radio signal. The antenna reflector captures incoming signals and concentrates those signals into a feed horn. The electronics associated with the antenna is "fed" the signal, then processes it electronically. For the transmission process the signal is generated as a carrier wave and sent through the feed horn to be transmitted by bouncing off of the concentrating antenna reflector. The antenna reflector is the most visible part of an antenna, but the entire package of the reflector, the feed system, electronics and power supply constitutes the entire device.

aperture dimension The diameter of a parabolic antenna is normally provided so that the area of the antenna reflector can be calculated.

APT The Asia Pacific Telecommunity This organization is headquartered in Bangkok, Thailand, and helps to coordinate telecommunications policy in Asia. Substantial support for this organization is provided by Japan.

apogee The high point of an orbit. The perigee represents the lowest point in the orbit.

ASIC This acronym stands for application specific integrated circuit. These high performance and greatly miniaturized integrated circuits and monolithic devices allow satellite handsets to encode and decode satellite signals.

azimuth This is the east-west angle needed to align a ground antenna satellite with a GEO satellite. One needs to also have the elevation angle (north-south) to achieve the desired pointing angle to point to a GEO satellite.

byte 8 bits of information. Unit used in computer memory. Eight bits of information is more or less equivalent to the amount of information needed to represent a single word when digitally encoded.

c This is the symbol that represents the speed of light, which is 300,000 km/second.

C-band This is an RF spectrum band used for satellite communications (The uplink is at 6 GHz and the downlink is at 4 GHz.)

J.N. Pelton, *Satellite Communications*, SpringerBriefs in Space Development,
DOI 10.1007/978-1-4614-1994-5, © Joseph N. Pelton 2012

CCIR Consultative Committee on International Radio. The standards-making part of the ITU that addresses radio and satellite communications.

CDMA Code Division Multiple Access. A digital multiplexing system.

CITEL The telecommunications standards-making and coordinating body within the Organization of American States.

coder/codec A coder encodes a digital signal. A codec is a coder/decoder.

dB A decibel. This is a logarithmic expression based on the power of 10. In this system 3 dB represents a doubling of power. 10 dB would represent 10 times more power. 20dB would represent 100 times more power. 30 dB would represent 1,000 times more power, and 100dB would represent 10,000,000,000 times more power.

dBW The ratio of the power relative to one watt expressed in decibels.

EHF This stands for the extremely high frequency band of spectrum. This band goes from 30 GHz up to 300 GHz.

EIRP Effective Isotropic Radiated Power.

elevation The upward tilt of an antenna to "see" a satellite; zero degrees is the horizon. The azimuth is the other orientation angle needed to align the antenna reflector with the satellite in the east to west direction.

encoding A process whereby a digital signal can be converted into a complex digital code so that more condensed information can be transmitted by using various types of encoding systems. This can be Reed-Solomon encoding, Viterbi encoding, Turbo coding or other systems.

ETSI The European Telecommunications Standards Institute.

FM Frequency Modulation.

FDMA Frequency Division Multiple Access. An analog multiplexing system.

gain Measure of antenna performance. Gain is expressed as a ratio of output power to input power. 1 represents the irradiation pattern of an omni-antenna that irradiates energy equally in all directions as a complete sphere. A higher gain antenna represents the degree to which a highly focused beam can be formed as a smaller three-dimensional sector carved out of a perfect sphere. A gain of 10 dB would be sector that represents 1/10 of sphere. A gain of 20dB would be sector that represent 1/100 of a sphere and 30 dB would represent 1/1,000 of a sphere.

GEO geosynchronous orbit.

GHz Representation for gigahertz. This is the equivalent of 1 billion cycles per second.

Hz This stands for a Hertz. A Hertz refers to cycles per second and is a measure of frequency.

IEC The International Electro-Technical Commission, headquartered in Paris, France.

IEEE The Institution of Electrical and Electronic Engineers.

IETF The Internet Engineering Task Force.

inclination The inclination of an orbit measured in degrees. 0 degrees inclination means that a satellite is exactly in the equatorial plane. 90 degrees inclination means that the satellite is exactly in a polar orbit moving in a north to south orbit perpendicular to the equator.

International Frequency Registration Board A part of the ITU where use of radio frequencies by member countries are filed

IP Internet Protocol.

ISDN Integrated Services Digital Network. A series of standards for digital communications.

ISL Inter Satellite Link. This is also sometimes called a crosslink.

ISO The International Standardization Organization of Geneva, Switzerland. Also known as the OSI (French initials).

ITU The International Telecommunication Union, a U. N. organization headquartered in Geneva, Switzerland.

Ka-band Frequency band used for satellite communications. Uplink is at 30 GHz and downlinks are at 20 GHz.

KHz This is the representation for kilohertz (1,000 cycles per second).

Ku-band Frequency band used for satellite communications. Uplink is at 14 GHz and downlink is at 12 GHz.

L-band The frequency band used for mobile satellite services typically at the high end of this band at 1,500 MHz to 1.600 MHz.

LEO low Earth orbit.

MEO medium Earth orbit.

MHz This is the representation for megahertz, or 1,000,000 cycles per second.

modulator The device that modulates a signal. A modem is a modulator and demodulator.

Mux/DeMux A multiplexer and a demultiplexer.

NASA The National Aeronautics and Space Administration of the United States.

omni antenna This is an antenna that can irradiate a signal in all directions with equal power. An omni antenna represents the reference point for all higher gain antennas that can concentrate a signal in a specific direction, like a radiated cone.

orbital characteristics Description of an particular orbit. The main orbits used in satellite communications include (LEOs) low earth orbit constellations, (MEOs) medium earth orbit constellations, (GEOs) geosynchronous earth orbits (also called Clarke Orbits) and highly elliptical orbits (HEOs).

perigee This is the lowest point in a satellite's orbit. It is the opposite of the apogee, which is the highest point.

PCM Pulse Code Modulation.

PM Phase Modulation.

polarization This is a process of splitting radio frequencies into two usable parts so that spectrum can be effectively re-used – in this case doubling the usable spectrum available for satellite communications. The polarization technique can be either orthogonal polarization (i.e., by splitting the waveform signal into a vertical orientation or a horizontal orientation) or by circular polarization (i.e., by splitting a left-hand circularized waveform from a right-hand circularized waveform). This splitting process can separate the "wanted signal" from the "unwanted signal" by as much as 20 dB to 30 dB. This is to say the two polarized signals are effectively distinguished and separated from each other by passing through the polarizing system that "filters" the two signals apart, just as "Polaroid" sunglasses separate "wanted" light from "unwanted" solar glare.

PSK Phased Shift Keying.

RF radio frequency.

RFC Request for Comment. The IETF seeks on an ongoing basis to develop guidelines for utilizing the Internet more effectively and thereby to allow operating process changes to achieve that end. When a new process for using the Internet is proposed then members of the Internet Engineering Task Force are sent formal "Requests for Comment" to either agree or suggest alternative approaches.

SHF This stands for the Super High Frequency band of spectrum. This band goes from 3,000 MHz (or 3 GHz) up to 30 GHz.

TDMA Time Division Multiple Access. A digital multiplexing system.

transponder The essential electronic component in a communications satellite that is responsible for receiving, filtering, amplifying, converting the uplink frequency to a downlink frequency and transmitting the downlink carrier wave back to Earth. Communications satellite transponders are typically divided into frequency ranges of 36 to 72 MHz, but can be devised and manufactured in larger spectrum bands.

UHF This stands for the ultra high frequency band of spectrum. This band goes from 300 MHz to 3000 MHz.

VHF This stands for very high frequency.

VoIP Voice over Internet Protocol.

WTO World Trade Organization. This organization replaces the former General Agreement on Tariffs and Trade.

Appendix 2: Selected Bibliography and Books for Further Reading

Berlin, Peter. *The Geostationary Applications Satellite*, (1988) Cambridge University Press, New York, NY.

Chartrand, Mark R. *Satellite Communications for the Non Specialist* (2004) SPIE, Bellingham, Washington, D. C.

Elbert, Bruce. *Introduction to Satellite Communication* (2008) Artech House, Boston, Massachusetts, USA

Gordon, Gary D., and Walter L. Morgan. *Principles of Communications Satellites* (1993) John Wiley and Sons, New York, NY.

Lewis, Geoffrey E. *Communications Services Via Satellite*, (1988) BSP Professional Books, Oxford, England, UK.

Lida, Takashi (editor). *Satellite Communications System and Its Design Technology* (2000) IOS Press, Tokyo, Japan.

Maral, Gerard, Michel Bousquet and Zhili Sun. *Satellite Communications Systems: Systems, Techniques, and Technology* (2010) 5th Edition, John Wiley and Sons, New York, NY.

Miya, Kenichi. *Satellite Communications Technology* (1985) 2nd Edition, KDD Engineering, Tokyo, Japan.

Pelton, Joseph N. *Basics of Satellite Communications* (2006) International Engineering Consortium, Chicago, Illinois, USA

_____ *Future Trends and Satellite Communications Markets* (2005) International Engineering Consortium, Chicago, Illinois.

_____ *Satellite Communications 2001: The Transition to Mass Consumer Markets, Technologies and Systems*, (2000) International Engineering Consortium, Chicago, Illinois.

Pratt, Timothy, and Charles W. Bostian. *Satellite Communications* (1986) John Wiley and Sons, New York, NY.

Pritchard, Wilbur L., Henri G. Suyderhoud, and Robert A. Nelson. *Satellite Communications Systems Engineering*, (1993) 2nd Edition, Prentice Hall Inc., Englewood, NJ.

Rees, David E. *Satellite Communications*, (1990) John Wiley and Sons, New York, NY.

Dennis S. Roddy, *Satellite Communications*, (2001) 3rd Edition, McGraw Hill, New York, NY.

Williamson, Mark. *The Communications Satellite*, (1990) Adam Hilger, Bristol, England.

Index

About the Author

Dr. Joseph N. Pelton is the award-winning author or editor of over 30 books and 300 articles in the field of space systems. His five-book series on technology and its impact on society includes: *MegaCrunch*, *e-Sphere*, *Future Talk*, *Future View*, and *Global Talk*, and he was nominated for a Pulitzer Prize for *Global Talk*. From 1992 to 1995 he served as Chairman of the Board and Vice President of Academic Programs and Dean of the International Space University of Strasbourg, France. He is currently a member of the ISU faculty and co-editor of two of their books: *The Farthest Shore* (2009) and *The Handbook of Satellite Applications* (2012). He is also the Director Emeritus of the Space and Advanced Communications Research Institute (SACRI) at George Washington University. This institute, which he headed from 2005 to 2009, conducted state-of-the-art research on advanced satellite system concepts and space systems. From 1988 to 1996 Dr. Pelton served as Director of the Interdisciplinary Telecommunications Program at the University of Colorado in Boulder, which at that time was the world's largest graduate level telecommunications program. Prior to that he held a number of positions at Intelsat and Comsat, including serving as Director of Strategic Policy and Director of Project Share for Intelsat.

Dr. Pelton is a Fellow of the International Association for the Advancement of Space Safety (IAASS), a member of its Executive Board and Chairman of its Academic Committee. He is the also Executive Editor of the IAASS publication series and acting President of the International Space Safety Foundation of the United States as well as the former President of the Global Legal Information Network. Dr. Pelton was the founder of the Arthur C. Clarke Foundation and remains as the Vice Chairman of its Board of Directors. This Foundation honors Sir Arthur Clarke, who first conceived of the communications satellite (in 1945). Dr. Pelton was elected to full membership in the International Academy of Astronautics in 1998 and was awarded in 2000 the Sir Arthur Clarke Award for lifetime achievement in the field of satellite communications. He was elected to the Hall of Fame of the Society of Satellite Professionals International in 2001, an honor only extended to some 50 people in the field. In 2004 he was elected an Associate Fellow of the American Institute of Aeronautics and Astronautics. His degrees in Physics and International Relations are from the University of Tulsa, New York University (NYU) and Georgetown University.